Multicomponent Transport in Polymer Systems for Controlled Release

Polymer Science and Engineering Monographs: A State-of-the-Art Tutorial Series

A series edited by **Eli M. Pearce,** Polytechnic University, Brooklyn, New York

Associate Editors

Guennadi E. Zaikov, Russian Academy of Sciences, Moscow
Yasunori Nishijima, Kyoto University, Japan

Volume 1
Fast Polymerization Processes
Karl S. Minsker and Alexandre Al. Berlin

Volume 2
Physical Properties of Polymers: Prediction and Control
Andrey A. Askadskii

Volume 3
Multicomponent Transport in Polymer Systems for Controlled Release
Alexandre Ya. Polishchuk and *Guennadi E. Zaikov*

In preparation

Quantum Chemical Aspects of Cationic Polymerization of Olefins
V.A. Babkin, Guennadi E. Zaikov and Karl S. Minsker

Multicomponent Transport in Polymer Systems for Controlled Release

Alexandre Ya. Polishchuk

and

Guennadi E. Zaikov

Russian Academy of Sciences
Moscow

Gordon and Breach Science Publishers

Australia • Canada • China • France • Germany • India • Japan • Luxembourg •
Malaysia • The Netherlands • Russia • Singapore • Switzerland •
Thailand • United Kingdom

Amsteldijk 166
1st Floor
1079 LH Amsterdam
The Netherlands

British Library Cataloguing in Publication Data

Polishchuk, Alexandre Ya.
  Multicomponent transport in polymer systems for controlled release.—(Polymer science and engineering monographs : a state-of-the-art tutorial series ; v. 3)
  1. Polymers 2. Polymerization
  I. Title II. Zaikov, G. E. (Guennadi Efremovich)
  547.7

ISBN 90-5699-594-4

CONTENTS

INTRODUCTION TO THE SERIES

This series will provide, in the form of single-topic volumes, state-of-the-art information in specific research areas of basic applied polymer science. Volumes may incorporate a brief history of the subject, its theoretical foundations, a thorough review of current practice and results, the relationship to allied areas, and a bibliography. Books in the series will act as authoritative references for the specialist, acquaint the non-specialist with the state of science in an allied area and the opportunity for application to his own work, and offer the student a convenient, accessible review that brings together diffuse information on a subject.

PREFACE

It is well known that the great ancient fabulist Aesop was a slave. Once, his master decided to go from point A to point B. He took with him heavy luggage and slaves to carry this luggage. Each slave had to transport one unit, and all the slaves tried to select relatively light packages. Only Aesop took the heaviest unit, the huge box with bread.

Everybody laughed, how stupid Aesop was. However, every day his box grew lighter and lighter. Time passed, all the bread was eaten, and there was nothing left for Aesop to carry. In the meantime, the weight of the other packages remained the same.

Writing this book, we also tried to look ahead more than to the past or the present. But this is our opinion. The opinions of the readers are much more important. We would like to hear from you. Please send your comments to:

Dr. Alexandre Ya. Polishchuk

      or

Professor Guennadi E. Zaikov

Institute of Biochemical Physics
United Institute of Chemical Physics
Russian Academy of Sciences
4 Kosygin Street, Moscow 117334
Russia

Tel.: 7095 9397-191/320
Fax: 7095 938-2156
e-mail: chembio@glas.apc.org

INTRODUCTION

Since the 1970s, the area of controlled release of drugs and other compounds has developed with high acceleration. The "Big Seven" countries have dominated three-fourths of the international pharmaceutical market [1], and leading pharmaceutical companies (Glaxo, Merck-US, BMS-US, Hoechst, Ciba, Sandoz, SB-UK/US, Bayer, and Koche-Switzerland) all over the world have become more and more involved in this area. The explanation can be found by looking at the benefits of controlled delivery of drugs. These benefits were formulated by Chien in one of the first-ever comprehensive books on this subject [2], as the following:

• Controlled administration of a therapeutic dose at a desirable delivery rate;

• Maintenance of drug concentration in the body at an optimal therapeutic level for prolonged duration of treatment;

• Minimization of the frequency of dose intake;

• Maximization of efficacy-dose relationship;

• Reduction of adverse side effects;

• Enhancement of patient compliance [3].

Chien also generalized the main physical approaches and corresponding technologies:

1. Controlled Drug Release by a Diffusion Process

   A. Membrane Permeation-Controlled Drug Delivery

      - non-porous membranes;

      - microporous membranes;

      - semipermeable membranes.

   B. Matrix Diffusion-Controlled Drug Delivery

      - lipophilic polymers;

      - hydrophilic polymers;

      - porous polymers.

C. Microreservoir Diffusion-Controlled Drug Delivery

- hydrophilic reservoir/lipophilic matrix;
- lipophilic reservoir/hydrophilic matrix.

2. Controlled Drug Release by an Activation Process

D. Osmotic Pressure-Activated Drug Delivery

E. Hydrodynamic Pressure-Activated Drug Delivery

F. Vapor Pressure-Activated Drug Delivery

G. Magnetism-Activated Drug Delivery

H. Ultrasound-Activated Drug Delivery

I. pH-Activated Drug Delivery

J. Ion-Activated Drug Delivery.

The decade has passed, but the above general principles are still valid. Moreover, Lee wrote in the same book [4] with a reference to one of his previous papers [5] that the kinetics of drug release from a matrix containing uniformly dissolved or dispersed drugs have already have been documented quite enough. Does this mean that there is nothing left to do in the theory of controlled release and its application in practice? We answer this question with a "No." The shape is the same, but the matter is remarkably changed. Ten years ago, phenomenological and empirical approaches predominated in almost all fields of controlled release, such as fibrous delivery systems [6], transdermal drug delivery [7], and diffusion-controlled release of drugs from coated polymer complexes [8], although the first attempts to do theoretical justification of these processes had already been made. Lee examined theoretically the effect of nonuniform distribution and considered five different expressions for initial distribution. The constant rate of release could be achieved via a specific sigmoidal drug concentration distribution without the need to have a saturated reservoir. Hydrogel polymers were named as particularly suitable for this application [4]. Later this idea was brilliantly confirmed [9].

Today, all serious discussion of the problem of controlled release almost inevitably leads to theoretical justification over the right choice of delivery system. A large part of this work has already been done and is summarized in the following books.

In 1987, Lee and Godd [10] considered novel approaches to the delivery of bioactive agents, including pharmaceutical, agricultural, and veterinary compounds. They discussed topics ranging from hydrophilic polymers, hydrogels, microspheres, and biodegradable polymers, to transdermal and transmucosal delivery systems, rate control by means of geometric configuration, and disposable controlled-release devices. Several models for diffusion in polymers and percutaneous absorption were also included.

El-Nokaly et al. [11] presented recent advances in the basic research and applications of polymeric delivery systems. They examined the preparation and properties of biodegradable and other polymer delivery systems, and stimulated new ideas for the application of polymeric drug delivery technology to cosmetics, foods, pharmaceuticals, and pesticide systems.

Dunn and Ottenbrite edited a book [12] covering a broad spectrum of methods of drug delivery. Focusing on the use of polymers and other materials, this volume presented new materials and methods for controlled drug release. Emphasizing selection of materials for drug delivery rather than just reviewing current methods, the editors challenged researchers to venture into new areas and consider new methods. They examined recent advances in biodegradable and bioerodible polymer matrices for drug delivery. In particular, an introductory section of four tutorial chapters ("Biologically Active Polymers," "Polymeric Matrices," "Liposomes," and "Interactions Between Polymeric Drug Delivery Systems and Biological Systems") was of great value to those just entering the field.

Shalaby and colleagues provided readers with fundamental information on the organic and physical chemistry of synthetic and natural polymers, as well as recent developments in polymer synthesis and modification. Opening with a section on theory and design of water-soluble polymer systems, their volume discusses basic properties as well as a range of contemporary applications. Thirty-three chapters divided into five sections cover polymers and intermediates, polymer synthesis and modification, physicochemical aspects of aqueous solutions, biomedical and industrial applications, and advances in less conventional systems [13].

Soviet scientists contributed to the theory of controlled release with books and reviews of representatives of the scientific groups of Plate [14, 15], Zaikov [16, 17], and Torchilin [18].

Now, any book dealing with technology forecast includes the theoretical basis [19]. The theory is presented in any communication considering the future of medical and engineering application of polymers [20, 21]. Any development of a new generation of polymers requires theoretical study of the regularities of multicomponent transport in such materials [22]. Significant theoretical opportunities and challenges exist in the creation and characterization of biomaterials [23].

Development of new drugs for known and novel diseases also requires theoretical study of their mobility in different media [24, 25]. One example: several approaches followed in the development of strategy for the treatment of HIV infection [20]. Theoretical analysis of the effect of drug delivery systems on the efficacy and toxicity might further improve the anti-HIV treatment.

Application of systems of controlled release has extended greatly during the last two decades and now covers medicine, agriculture, the chemical industry, and other fields of human activity. For this reason, it would be impossible to describe all theoretical approaches to the processes of release of different compounds from different polymeric systems. We selected only one field: the theory of multicomponent transport. This seems to us the basic point for the further development of systems of controlled release—possibly due to our special interest in the field. In this book we do our best to show how the theory can or could be applied in practice. The reader will have to estimate the result of our attempt.

REFERENCES

1. *Med. Ad. News,* **5** (1994).
2. C.G. Gobelein, C.E. Carraher, Jr. (Eds.), *Polymeric Materials in Medication.* Plennum Press, New York & London (1984).
3. Y.W. Chien, In: *Polymeric Materials in Medication.* C.G. Gobelein, C.E. Carraher, Jr. (Eds.), p. 27, Plennum Press, New York & London (1984).
4. P.I. Lee, *Ibid,* p. 79.
5. P.I. Lee, *J. Memb. Sci.,* **7,** 225 (1980).
6. R.L. Dunn, J.W. Gibson, B.H. Perkins, In: *Polymeric Materials in Medication.* C.G. Gobelein, C.E. Carraher, Jr. (Eds.), p. 47, Plennum Press, New York & London (1984).
7. J.H. Gaskill, P.P. Sarpotdar and R.P. Giannini, *Ibid,* p. 61.
8. Y. Raghunathan, L. Amsel, O. Hinsvark, K. Rotenberg, *Ibid,* p. 73.
9. M.V. Badiger, M.E. McNeill, N.B. Graham, *Biomaterials,* **14,** 1059 (1993).
10. P.I. Lee, W.R. Good (Eds.), *Controlled-Release Technology: Pharmaceutical Applications,* ACS Symposium Series, No. 348 (1987).
11. M.A. El-Nokaly, D.M. Piatt, B.A. Charpentier (Eds.), *Polymeric Delivery Systems: Properties and Applications,* ACS Symposium Series, No. 520 (1993).
12. R.L. Dunn, R.M. Ottenbrite (Eds.), *Polymeric Drugs and Drug Delivery Systems,* ACS Symposium Series, No. 469 (1991).
13. Sh. W. Shalaby, Ch. L. McCormick, G.B. Butler, (Eds.), *Water-Soluble Polymers: Synthesis, Solution Properties, and Applications,* ACS Symposium Series, No. 467 (1991).
14. N.A. Plate, L.I. Valuev, *Polymer Contact with the Living Body,* Issue 8, Znanie, Moscow (1987). (In Russian.)
15. N.A. Plate, A.E. Vasilyev, *Physiology of Active Polymers,* Khimiya, Moscow (1986).
16. G.E. Zaikov, V.S. Livshits, In: *Polymer Yearbook,* R.A. Pethrick and G.E. Zaikov (Eds.), p. 177, Harwood Academic Publ., New York (1987).
17. A.L. Iordanskii, T.E. Rudakova, G.E. Zaikov, *Interaction of Polymers with Bioactive and Corrosive Media,* VSP, Utrecht, The Netherlands (1994).

18. V.V. Torchilin, A.L. Klibanov, V.N. Smirnov, In: *Liposomes and Interaction with Cells,* Moscow University Press (1981).
19. M. Gillespie, *3i Technology Forecast,* Issue Number 2, *Drug delivery systems into 1990s,* 3i Group Plc. (1990).
20. F. Kochinke, J.J. Vajda, V.G. Wong, *Investigative Ophthalmology and Visual Science,* **35,** 2219 (1994).
21 R. Baker, J. Colvin, *Journal of Biomedical Engineering,* **13,** 530 (1991).
22. L. Brannon-Peppas, *Polymer News,* **20,** 20 (1994).
23. N.A. Peppas, R. Lamger, *Science,* **263,** 1715 (1994).
24. M. Stewart, *Clin-Chem.,* **40,** 953 (1994).
25. H. Mirchandani, Y.W. Chien, *International Journal of Pharmaceutics,* **95,** 1 (1993).

1 GENERAL ASPECTS OF TRANSPORT OF LOW-MOLECULAR WEIGHT COMPOUNDS IN POLYMERS

1.1. THEORY OF WATER SORPTION IN POLYMERS

When we discuss the regularities of diffusion of low-molecular weight compounds we primarily talk about the diffusion of aqueous solutions of acids, bases, and salts. This is because of the remarkable role which water plays in biology, medicine, agriculture, and industry. Water is the most common solvent for electrolytes and, therefore, the mechanism of their release depends on water-polymer interactions. With respect to water and aqueous solutions, polymers have been classified as hydrophobic, hydrophilic, and moderately hydrophilic (moderately hydrophobic) [1, 2]. Water sorption in polymers possesses all features which are general for sorption of other solvents. In the meantime, there are many specific features which characterize water sorption only. Below we consider the most important of them.

1.1.1. Water sorption in hydrophobic polymers

Polyolefins, fluoroplastics, rubbers, polysiloxanes, polyethers and polyesters, and polystyrene and polyvinylchloride are typical examples of hydrophobic polymers [3]. Under normal conditions they contain less than 1% of water in equilibrium. Therefore, the majority of polymers widely used in various fields of industry, agriculture, and medicine [4, 5] should be considered as hydrophobic polymers.

1

The development of the theory of water sorption in hydrophobic polymers faces some difficulties because experimental investigations are often performed with commercial samples. These samples may contain a dozen hydrophilic additives (catalysts, monomers, stabilizers, etc.) The effect of fillers and additives on water sorption in hydrophobic polymers will be considered in more detail in chapter 2. Here we focus our attention on the influence of chemical structure, polymer morphology, and water association on the behavior of water-hydrophobic polymer systems with references to the recent theoretical and experimental achievements in this area.

The water sorption isotherm for hydrophobic polymers at room temperature undergoes Henry law or shows deviation above linearity [6, 7]. It should be mentioned that the experimental value of solubility sometimes depends on the method of measurement [7]. For interpretation of the nonlinear character of water sorption, the following models are generally used: (i) model of multilayer BET adsorption [8], (ii) Flory theory of polymer solution [9], and (iii) model of cluster formation by Zimm and Lundberg [10]. It is important to distinguish the formation of clusters in the binary system of polymer and water from the formation of water associates occurring around hydrophobic admixtures and inculcation groups. The real polymer can contain (i) homogeneous dissoluble water existing as separate molecules, (ii) water clusters, (iii) hydrated water of polymer functional groups (inculcation groups), and (iv) water immobilized on polymer admixtures. As water activity varies one of these mechanisms may predominate, simplifying the modeling of sorption.

Neither BET model nor Flory-Huggins equation can be applied for a case of cluster formation in polymers. The first has no physical sense for nonpore polymer system, and the second does not take into account a dozen water–water contacts. Zimm and Lundberg [11] proposed the equation for the formation of clusters in polymers which was based on statistical thermodynamics of the binary system.

$$G_w/V_w = (1-\phi_w)(d(a_w/\phi_w)/dt + 1)$$

where ϕ_w is the water volume portion in the polymer, a_w is its activity in liquid phase, V_w is the water molar volume, and G_w is the integral of a cluster formation. The average number of molecules N_c in the cluster follows the expression [6]

$$N_c = \phi_w (1 + G_w/V_w) + 1.$$

For a case of uniform water distribution in the polymer, $G_w/V_w = -1$. As this ratio exceeds -1, clusters should be observed in the system.

The Barrie-Platt model [12] considered water as a tetrafunctional monomer. This is one of the most important among other models of the formation of clusters [6]. The process of cluster formation itself was simulated by quasipolymerization which led to the formation of a hydrogen bond net of water molecules included in the cluster.

Water sorption isotherm in nonpolar and low polar polymers may be described in terms of the model of cluster formation [6]. The process of water association in clusters may be presented by a number of consecutive reactions

$$w + w = w_1 \quad (K_1)$$

$$w + w_1 = w_2 \quad (K_2)$$

$$\dots\dots\dots \qquad \dots \tag{A}$$

$$w + w_{n-1} = w_n \quad (K_n)$$

where w represents a free water molecule, and w_1, w_2, ... w_n are water clusters with 2, 3, ..., n+1 molecules. The corresponding equilibrium constants are shown in parentheses. Following (A) the total water concentration in the polymer is given by

$$C_w = C_{wf} + 2C_{w1} + 3C_{w2} + \cdots + (n+1)C_{wn} =$$

$$= C_{wf} + 2K_1C^2_{wf} + 3\,K_1K_2C^3_{wf} + \cdots + (n+1)K_1K_2\cdots K_nC^{n+1}_{wf}$$

or in form

$$C_w = C_{wf}\,[1+\sum_{i=1}^{n} (i+1)(\prod_{j=1}^{i} K_j)C^i_{wf}] \tag{1.1}$$

where C_w and C_{wf} are the concentrations of total and free water in polymer, respectively, and $\Pi\,K_j$ is the product of i constants. The Henry law

$$C_w = C_{wf} = K_Ha_w \tag{1.2}$$

is the particular case of the equation (1.1). A cluster may also be formed due to the reaction $w_i + w_j = w_{ij}$, where i, j > 2, but such aggregates usually do not contribute to water diffusion because of much lower mobility than free molecules. Generally, three types of water isotherms are observed in hydrophobic polymers.

(i) The linear isotherms in the whole interval of water vapor activities according to Eq. 1.2. These isotherms are observed in high density poly-

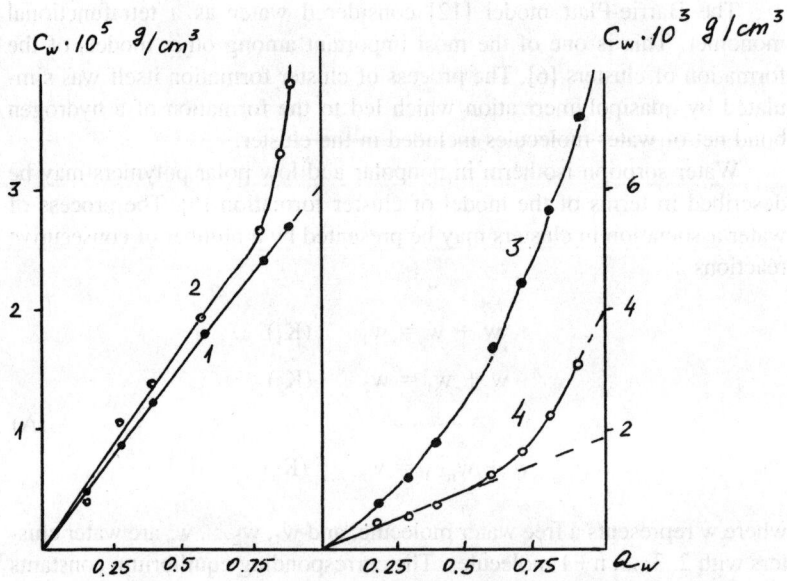

Figure 1.1. Isotherms typical for water sorption in hydrophobic polymers. 1-HDPE, 2- LDPE, 3- copolymer of ethylene with vinylacetate, 4- fluoroplastic F-32.

ethylene (HDPE), isotactic polypropelene (PP) [13], and polyethyleneterephthalate (PETP) [14].

(ii) The isotherms with deviation above linearity (concave curve) at all values of a_w. These isotherms were observed for polyvinylchloride (PVC) [7], polyethylenemethacrylate [15] and some other slightly polar polymers [6].

(iii) The isotherms with positive deviation at high activities only. Examples of such polymers are fluoroplastics [16], polybutyleneadiurethane [17], and poly(dimethylsiolane) (PDMS) [18].

The isotherms of all the above types are shown in Fig. 1.1.

The Henry law may be the consequences of two different mechanisms:

(a) formation of a true water solution in the hydrophilic matrix in accordance with Flory-Huggins equation

$$\phi_w = a_w \exp\left[-(1+\chi)\right]$$

where χ is the solubility parameter; and

(b) water adsorption on oxygen-containing groups

$$K_H = \exp\left[-(1+\chi)\right].$$

Obviously, the Henry isotherms might be considered as part of the Langmuir isotherm, but there are two reasons to not consider this case here:

(i) it is valid only for part of the kinetic curve;

(ii) the effect of hydrophilic fillers, additives, and impurities which cause the Langmuir will be described in chapter 2 in detail.

In the meantime, the reasons for deviation above Henry isotherms require some essential comments.

The analysis of Eq. 1.1 [16, 6] showed that when $a_w < a_{wf}$ we get for any i less than some number of l

$$(i+1)(\prod_{j=1}^{i} K_j)a_{wf}^i < \varepsilon \qquad (1.3)$$

where ε is the absolute error of measurement, and a_{wf} is the maximal activity of water vapor, for which the isotherm is linear. Eq.1.3 corresponds to the event when the contribution of clusters in total water adsorption is negligible in comparison with monomeric water sorption.

If for $a_w > a_l > a_{wf}$ the deviation above Henry's law is observed then for any i more than l

$$(i+1)(\prod_{j=1}^{i} K_j)a_{wf}^i < \varepsilon. \qquad (1.4)$$

The value of l which defines the validity of Eqs. 1.3 and 1.4 may be calculated from the expression

$$l > \ln\delta / \ln(a_{wf}/a_l)$$

where δ is the relative experimental error.

The analysis of the experimental results for water sorption in typical hydrophobic polymers (low density polyethylene (LDPE), fluoroplastics) [6, 16] showed that among all possible water associates, only certain N-mers contribute in sorption and diffusion. For this case Eq. 1.1 may be simplified to so-called bimodal distribution of water molecules

$$C_w = K_H a_w [1 + N(\prod_{j=1}^{N} K_j)(K_H a_w)^{N-1}] \qquad (1.5)$$

The solution of Eq. 1.5 allows to find the effective constant of sorption $G_N = \Pi\, K_j$ and the number of molecules in a cluster (N). The average number of molecules in the cluster may be found from 1.1, in general

$$N_c = C_{wf} / (C_{wf} + \sum_i^n C_{wi}) = \sum^n iG_N C_{wf}^i / \sum G_N C_{wf}^i.$$

For a case of bimodal distribution Eq. 1.5 transforms into simpler form.

$$N_c = 1 + (N-1)G_N(K_H a_w)^{N-1}/[1 + G_N(K_H a_w)^{N-1}]$$

If there is no possibility of ignoring any items in Eq. 1.1 the number of molecules in cluster and thermodynamic constants may be found through solution of higher order differential equation

$$d^n C_w / da_w^n = f(N, K_H, C_N)$$

or due to the certain physical limits concerning scheme (A).

For example, the model of ligand (or sorption) centers assumed the expression of thermodynamic constants K_i in system (A) through the single immobilization constant K_T [6].

$$K_i = K_T(N-i)/i$$

where the number of adsorption centers corresponds to the number of molecules in the associate. Eq. 1.1 transforms, respectively, to

$$C_w = C_{wf}(1 + K_T C_{wf})^N(1 + NK_T C_{wf})/(1 + K_T C_{wf}).$$

At $K_T C_{wf} \ll 1$ and $NK_T C_{wf} \ll 1$ we get the Henry law (1.2), and at $K_T C_{wf} \gg 1$ we get the expression

$$C_w = (N + 1)(K_T)^N C^{N+1}_{wf}$$

which described the certain type of sorption isotherm illustrated in Fig. 1.1. The examples of such isotherms were found for PVC and copolymers of vinylacetate with ethylene (PEVA).

The above model of a cluster formation allows us to calculate the main parameters of water sorption in a dozen hydrophobic polymers. The results of these calculations are presented in Table 1.1.

Table 1.1 shows the change of constant K_H as polymer varies. This

Table 1.1. Parameters of water sorption and diffusion in hydrophobic polymers [6, 16].

Polymer	$C_w \times 10^3$ g/g	$K_H \times 10^3$ g/g	ΔH kcal mol^{-1}	$D_{wf} \times 10^9$ cm^2 s^{-1}	$D_{wf} \times 10^9$ cm^2 s^{-1}
HDPE	0.036	0.036	5.9	2.3	—
LDPE	0.051	0.041	6.2	2.6	>0.5
F-10	0.80	0.80	8.5	1.5	—
F-2M	2.1	0.70	5.7	3.4	2.2
F-32	4.4	1.2	8.5	1.3	2.5
PE/VA 6:1	10.5	2.6	—	1.4	0.92
PE/VA 3:1	16.0	4.1	—	0.65	1.0
PVC	8.1	1.2	5.1	7.5	1.6

constant is affected by the nature of material, prehistory, and degree of the degradation during the processes of production and exploitation. It shows the effect of structure on sorption and diffusion processes in polymer. Unfortunately, many authors mention the linear character of water sorption in hydrophobic polymers without explanation of the mechanism of this phenomena.

In the meantime, the little change in the average number of water molecules in the cluster [6, 16] seems understandable due to the fact that water–water interaction predominates in a hydrophobic matrix. This is of course in the absence of hydrophilic additives or impurities, the effect of which is remarkable for natural and synthetic rubbers.

Sorption and transport properties of water through films of Nylon-6 were obtained at 5, 23, and 40°C. Commercially available films were used, and a Cahn electrobalance was employed for measuring the water uptake by the polymer samples. Values of the water sorption isotherms were accurately described by the Langmuir/Flory-Huggins dual-mode sorption model. At water activity values below 0.15, the volume fraction of water described by the Langmuir portion of the model was greater than the Flory-Huggins population. Solubility and diffusion coefficients of water, as well as the diffusion activation energy and enthalpy of dissolution of water for Nylon-6, were determined from the sorption experiments. Values obtained support the hypothesis of a bimodal water sorption mode, and the formation of water clusters [19].

The mechanism of water sorption may be affected by temperature. The diffusive and mechanical behavior of tetraglycidyl diaminodiphenyl methane (TGDDM) resin-based composites and diglycidyl ether of bisphenol-A (DGEBA) resin-based graphite/epoxy composites were investigated during water sorption at different temperatures. The water-absorption kinetics in both systems at 50, 70, 90, and 100°C were fitted by a Fickian diffusion model. However, a Langmuir-type, two-step sorption behavior was observed for water transport in DGEBA-based systems at 50 and 70°C. Using scanning electron microscopy, internal cracks due to water-absorption were found in the DGEBA-based samples after conditioning at 90 and 100°C in water, whereas no cracks were detected in TGDDM-based samples conditioned in water at 100°C. Ultrasonic testing did not show significant change of modulus or density in the TGDDM-based samples conditioned in water at 100°C. No significant changes of dynamic modulus or damping factor were observed for the TGDDM-based samples redried after immersion in 100°C water, whereas slight changes were observed above 120°C for the samples containing absorbed water. However, both water-containing and redried DGEBA-based samples showed a decrease of dynamic modulus and an omega-transition around 120°C. A single-fiber fragment test revealed that the absorbed water at 80°C reduced significantly the interfacial shear strength of DGEBA/DDA resin-AS4 fiber samples and DGEBA/DDA resin-AU4 fiber samples [20].

The results of macroscopic study were recently confirmed by the method of computer simulation. Molecular dynamics simulations were carried out in order to examine the mechanism of diffusion of small penetrants in amorphous polymer membranes. Diffusion processes of methane, water, and ethanol in PDMS and in polyethylene (PE) were investigated. Pure liquid water was also simulated. The insertion probabilities $P(R)$ of hard-sphere atoms of radius R into the polymers and the liquids were calculated. The free volume fraction, $P(0)$, of PDMS was large and the insertion probability of a finite size atom into PDMS was widely distributed compared with the other polymers and liquids. Simulations of 5 ns were performed for PDMS and in PE. Aggregates of water were found to be formed in PDMS. Diffusion coefficients of water in PDMS were reduced by more than 1 order of magnitude due to the aggregation. The calculated diffusion coefficients of the nonaggregated penetrants in PDMS and in the pure liquids agree well with the experimental values [21].

1.1.2. Water sorption in hydrophilic polymers

The most polymers which found application in systems of controlled release are hydrophilic or moderately hydrophilic. They take more than 10% and between 1 and 10%, respectively. Let us start this section with one example which shows the importance of theoretical study of water absorption in hydrophilic polymers.

The dependence of the long-term stability of a soft denture liner on the sorption and solubility of the liner was investigated. Because sorption and solubility are accompanied by a volumetric change, bacterial infestation, hardening, and color change, it is a physical property of importance. The purpose of that investigation was to determine the sorption and solubility of 12 soft denture liners (Verno-Soft, Super Soft, ProTech, Soft-Pak, Flexor, Novus, Molloplast-B, Durosoft, Justi Soft, Velvesoft, VinaSoft, and Prolastic). They include nine copolymers, two silicones, and one polyphosphazene fluoroelastomer. The sorption and solubility test was performed as outlined in American Dental Association (ADA) specification 12 for denture base polymers. Five specimens of each material were tested and data were collected at 1 week, 1 month, 3 months, 6 months, and 1 year. Sorption data varied from 0.2 to 5.6 mg/cm^2 at 1 week; 0.3 to 12.5 mg/cm^2 at 1 month; 0.1 to 22.0 mg/cm^2 at 3 months; 0.1 to 13.6 mg/cm^2 at 6 months; and 0.1 to 35.7 mg/cm^2 at 12 months. Solubility data varied from 0.0 to 0.4 mg/cm^2 at 1 week; 0.1 to 0.8 mg/cm^2 at 1 month; 0.1 to 1.2 mg/cm^2 at 3 months; 0.0 to 1.9 mg/cm^2 at 6 months; and 0.2 to 2.3 mg/cm^2 at 1 year. A statistical analysis of the data by two-way ANOVA and calculated Tukey intervals showed significant differences between materials at all time intervals. The results of this study have clinical implications because the sorption and solubility may affect the long-term life expectancy of the soft denture liner [22].

The isotherms of water sorption in hydrophilic polymers differ greatly from those for hydrophobic ones. S-form of the curves (Fig. 1.2) at low and moderate water concentrations may be explained by the strong interaction of water molecules with highly polar polymers. For highly elastic polymers, sorption isotherms are described by the van Goff equation. A typical example of such a behavior is shown for example in the lysozyme-water system [23].

Alteration of hydrophilic character of polymer matrix by controlled degradation [24] gave us an opportunity to see how the thermodynamics of water sorption varies as hydrophilic character of polymer changes. This was observed with initially water soluble and then partly degraded copolymer of maleic anhydride and vinyl acetate.

The first stage of interaction of the copolymer with water vapor was characterized by reversible sorption up to the equilibrium value. The shape of the curve remained convex as in Langmuir adsorption. Increase of the activity of water vapors caused irreversible binding of the sorbed water, rupture of chemical bonds in the copolymer with change in its structure (hydrolysis), and at last dissolution. Respectively, the sorption curve went up sharply. The critical concentration for hydrolysis was only reached in sorption at activities close to 1 (liquid phase). The general pattern of water uptake by partially degraded (less hydrophilic) copolymer was the same

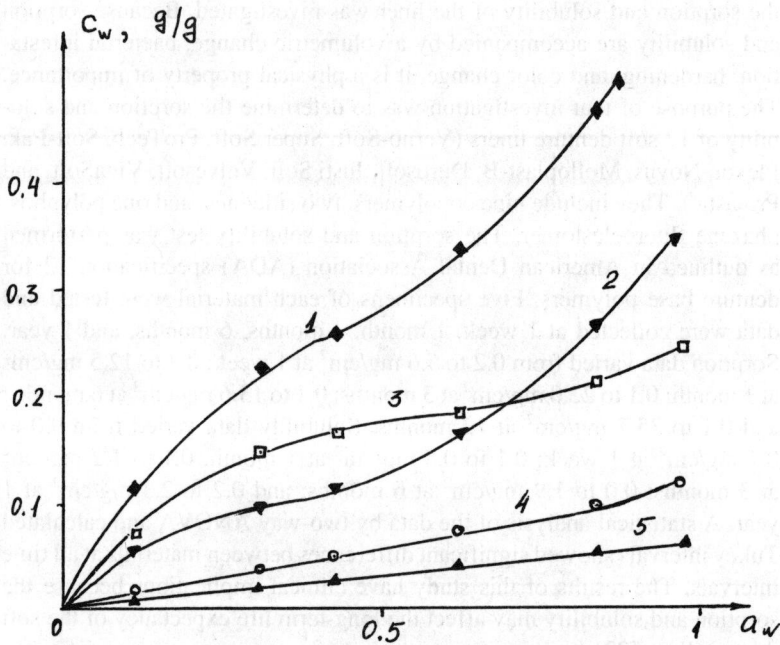

Figure 1.2. Water sorption isotherms of hydrophilic collagen (1), PVA (2), keratin (3), and moderately hydrophilic cellulose acetate (4) and Nylon 6,6 (5).

as for the original material, and the differences were in the depth of the structural changes following the point of equilibrium sorption. Respectively, the first (Langmuir) part of the curve became longer. Degradation for 30 minutes gave moderately swelling product in which the stage of irreversible sorption was restricted. The isotherms of water sorption indicated the different hydrophilic character of the samples and showed how the hydrophilic character of matrix varied from water soluble (initial copolymer) to low-swelling polymer (40 minutes of the degradation) and finally to hydrophobic matrix with Henry's law isotherm (Fig. 1.3).

The water uptake of several perfluorosulfonic acid membranes from liquid water over the temperature range 25 to 130°C and from water vapor at 80°C was studied in [25]. For water sorption from liquid water, the water uptake depended on the immersion temperature, the ion exchange capacity of the membrane, and the pretreatment of the membrane. The effects of pretreatment were not significant at immersion temperatures higher than 100 to 110°C. For water uptake from water vapor at 80°C, the sorption isotherms were similar in shape to those reported by previous inves-

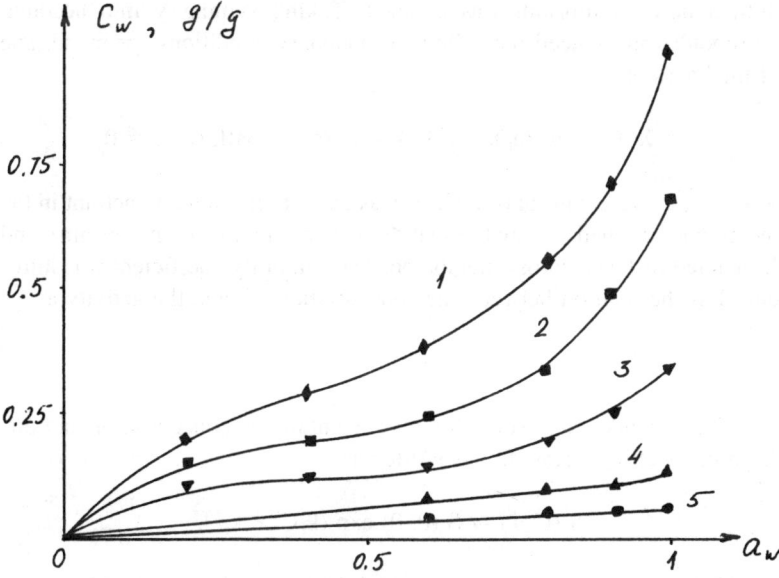

Figure 1.3. Isotherms for the sorption of water by a MAn/VAc copolymer (1) and its residues after thermal degradation at 10 (2), 20 (3), 30 (4), and 40 (5) minutes.

tigators for 18.5 to 30°C [6], although the water uptake at 80°C was less than that reported for the lower temperatures.

The load of additives is another possibility to alter the mechanism of water sorption. Lee investigated the effect of hydrogen peroxide on the physicochemical properties of semicrystalline poly (vinyl alcohol) (PVA) [26]. Significant and irreversible increases in equilibrium water swelling, dissolved oxygen permeability, and surface wettability had been observed in PVA samples treated with concentrated hydrogen peroxide. Based on the small amount of carbonyl content detected and the crystallinity reduction in hydrogen-peroxide-treated PVA samples, a mechanism involving a combination of hydrogen-peroxide-induced oxidative chain scission and dissolution of crystalline regions in PVA was proposed for the observed swelling kinetics and associated changes in polymer properties [26].

The effect of relaxation and swelling processes on thermodynamic solubility coefficient was studied theoretically by Petropoulos [27]. He considered an adsorption (desorption) process, in which the membrane was preequilibrated with penetrant vapor at activity a_1 (a_0) and then, at time $t = 0$ the surrounding penetrant vapor activity was changed simultaneously to a higher (lower) value a_0 (a_1) which was maintained constant thereafter until a new equilibrium was attained. Taking symmetry into account, Petropoulos introduced the following boundary conditions for membrane of thickness of 2l.

$$a(X, 0) = a_1 (a_0); \quad a(0, t) = a_0 (a_1); \quad \delta a(l, t)/\delta x = 0$$

where a_0, a_1 were constants, a (X, t) was the activity of the penetrant in the membrane (defined as equal to that in the vapor phase at equilibrium), and X denoted distance across membrane. The solubility coefficient was introduced as the relation between the concentration, C, and the activity a

$$S = C/a.$$

The dependence of solubility coefficient on the concentration and differential swelling stress, f, was written as

$$S (C, f) = S (0, 0) \exp (k_c C + k_f f).$$

The effect of stress-dependent solubility coefficient on the kinetics of sorption of solvent is shown in Fig. 1.4.

The mechanism of water sorption by hydrophilic polymers is essential for preparation of different polymeric systems. Permeability of oxygen and nitrogen gases was measured for crosslinked PVA and poly-4-vinylpyridine membranes swollen with aqueous solutions of nonvolatile electrolytes and nonelectrolytes. It was found that the permeability of the two gases change characteristically with the water content of the membranes, giving the ratio in the range 25 to 0.5 at around 0.3 of the water content. The importance of the hydration layers of the substrate polymers as the permselective field is suggested based on the measurements of DSC and the partition of the nonvolatile compounds for the swollen membranes. Diffusion selectivity and selectivity in solubility comprising the permselectivity were discussed [28].

The barrier characteristics of polymeric packaging materials were measured by the intensity of the molecular exchange between the packaged product and its external environment. The exchange of penetrant molecules through polymeric materials was determined by the capacity of a polymer matrix to sorb the penetrant molecules, and the ability of the penetrant to diffuse through the polymeric material according to Fick's laws. However, the presence of other small relative molecular mass compounds in the polymer matrix might change the way in which the permeant is sorbed and diffused. This case is such when water molecules were present in hydrophilic polymers affecting barrier characteristics of those polymers to penetrants

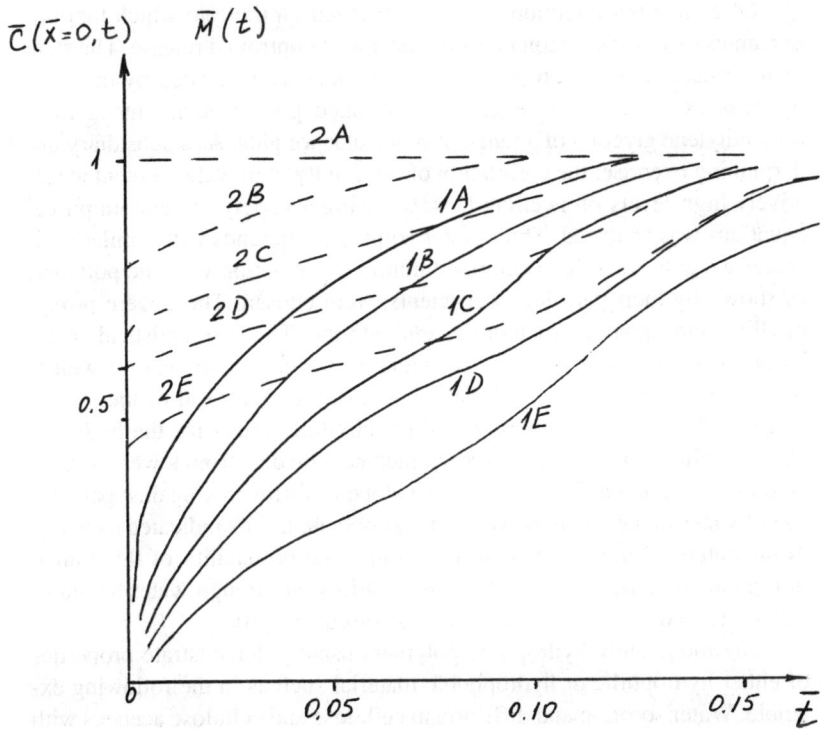

Figure 1.4. Computed absorption (1) and surface concentration (2) kinetic curves. Effect of the degree of stress dependence of S: S(C, f_{max})/S(C(a_1,), 0) = 1.6 (B), 2.5 (C), 5 (D), 10 (E), A = "Fickian curve."

such as oxygen or organic compounds. The oxygen barrier properties of hydrophilic polyamides change as a function of the amount of water in the polymer matrix. In addition, polymer morphology plays an important role in controlling the mode in which water molecules affect the transport of the penetrant. The change in oxygen barrier properties is then a function of the polymer water activity and morphology. A dual mode sorption model based on the Flory-Huggins and Langmuir equations was applied to the sorption of water by an amorphous and a semicrystalline polyamide, at three temperatures. The model provided the basis for the interpretation of the effect of water on the permeability, solubility, and diffusion of oxygen within the polyamides. Different interaction mechanisms of water and oxygen within the polymers, to include self-association of water molecules and oxygen-water molecular competition, were discussed at different water activity values for both polyamides [29].

Of course our attention should be focused on the role which thermodynamics of water sorption plays in systems of controlled release. The state of water-association with poly(ethylene oxide), as evidenced by diffusivity, was examined in a series of crosslinked polyurethanes made from poly(ethylene glycols) of a range of molecular weights. As a subsidiary underpinning exercise, the correlation of diffusivity with water content at relatively high levels of swelling (>45%) using a variety of semi-empirical equations was analyzed. Three water-soluble compounds with similar molecular weights and which exhibited minimal interaction with the polymer, as shown by their partition coefficients, were chosen. These were proxyphylline, morphine hydrochloride, and caffeine. The best statistical correlation of the data was obtained for plots of: (a) diffusivity against weight percent water; and (b) log diffusivity against the reciprocal of the weight percent of water in the hydrogels. Proxyphylline results for the high levels of swelling compositions were augmented with data from lower swelling compositions, and a clear break in the slope of diffusivity against percentage of water in the swollen hydrogel was obtained. This indicated a change in the nature of the diffusion at this point. The probability of this transition point corresponding to a change for diffusion through water bound as trihydrate to diffusion in free water was discussed [30].

The moderately hydrophilic polymers usually demonstrate properties of either hydrophilic or hydrophobic material such as in the following example. Water sorption and diffusion in cellulose and cellulose acetates with different degree of substitution were studied in [31]. Thermodynamic parameters of polymer mixing with water were calculated, and characteristics of clustering and diffusion coefficients were estimated. Water diffusion in cellulose and water-soluble cellulose acetates was shown to be primarily controlled by thermodynamic affinity associated with hydrophilic hydration as well as by structural changes as induced by sorption. As the degree of substitution of cellulose acetates increased, thermodynamic affinity to water decreased. Water clustering was shown to be a main reason of decreased diffusion coefficients.

The thermodynamics of water sorption is very difficult to separate from its kinetics, in other words from diffusion of water. For this reason we shall return to this problem in the following paragraphs.

1.2. WATER DIFFUSION IN POLYMERS

1.2.1. Water diffusion in hydrophobic polymers

The transport of water molecules in hydrophobic polymers is a subject of both technical and commercial importance. In the case of polyolefins it is technically important because the factors, such as morphology and molecular interactions, that control diffusion are not well determined. Also, the existing theory of diffusion, based on macroscopic parameters, is difficult to interpret in terms of the interactions occurring between the polymer matrix and the diffusants. It is commercially important because the use of polyolefins in applications such as wire coatings, barrier layers, and structures is extremely widespread. Failure of such applications can impact the lifetime of large amounts of capital investment. Examples such as wire and cable, landfill liners, and automotive components all have expected lifetimes. [32].

The formation of water associates and interaction of water with polymer groups lead to visible decrease in the mobility of water molecules. For a case of full immobilization (water associates are immobile), the total flux of water is proportional to the gradient of the concentration of free water

$$J_w = -D_{wf}\, \text{grad}\, C_{wf}. \tag{1.6}$$

If we rewrite Eq. 1.6 in the form of Fick's law

$$J_w = -D_w\, \text{grad}\, C_w$$

we get

$$D_w = D_{wf}\, \delta C_{wf}/\delta C_w.$$

In other words, the effective diffusion coefficient D_w depends on the type of sorption isotherm. For example, Langmuir isotherm leads to the dependence

$$dC_{wf}/\delta C_w = 1 + S_{pf}K_L/K_H$$

and

$$D_w^{-1} = D_{wf}^{-1}(1+S_{pf}K_L/K_H). \tag{1.7}$$

The latter dependence was confirmed for a dozen PE samples with different water contents.

Eq. 1.7 is also in agreement with empirical dependence which was obtained in [5] and explained by the model of dual sorption.

However, the model of total immobilization is not valid when water associates contribute to the total flux of water. In this case, the water flux in the polymer can be written as

$$J_w = -\sum_i^N iD_i\, \text{grad}\, C_{wi} \tag{1.8}$$

where D_i is the partial diffusion coefficient of water associates with 1,2,...,N molecules. C_{wi} is the concentration of corresponding associate.

Eq. 1.8 is written in agreement with above scheme (A). Further transformation of this equation leads to the dependence of water flux on activity of water in external phase.

$$J_w = -\sum_i^N iD_i \, G_i \, (K_H grad \, a_w)^i.$$

The integral (effective) diffusion coefficient can be written in a generalized form

$$D_w = \sum_i^N iD_i \, G_i \, (K_H a_w)^i / \sum_i^N i \, G_i \, (K_H a_w)^i$$

or simplified for a case of movement of single molecules

$$J_w = -D_{wf} K_H grad \, a_w.$$

For a case of bimodal distribution of water in hydrophobic polymer which was reported in Section1.1.1, the total flux may be written as

$$J_w = -D_{wf} K_H grad \, a_w - NG_N D_N (K_H grad \, a_w)^N$$

where D_N is N-mer diffusion coefficient, and the number of molecules in a moving cluster is unknown.

The bimodal distribution and corresponding regularities of water diffusion were found for a number of fluoroplastics [16]. Then D_{wf} value was calculated for low activities when multimer formation might be ignored. Values of the corresponding constants K_H and G_N were defined from sorption experiments. This gave an opportunity to define unknown parameters D_N and N. The simplest way for such a calculation was the graphical analysis of the $J_w(a_w)$ dependence. This dependence is shown in Fig. 1.5 for all above fluoroplastics. The good agreement between experimental and theoretical results confirmed the validity of the model of partial immobilization.

The ligand model of Tanford [6] requires another approach to the description of water transport. For this case the permeability of water (P_w) may be written in form

$$P_w = P_{wf} (1 + K_T a_w)^{N-1}(1 + NK_T a_w) \tag{1.9}$$

where P_{wf} is the permeability of free water ($P_{wf} = K_H a_w \, D_w$), $K_T = K_{HT} exp \, (-V_w/BV_f)$ that is in agreement with free volume theory [33].

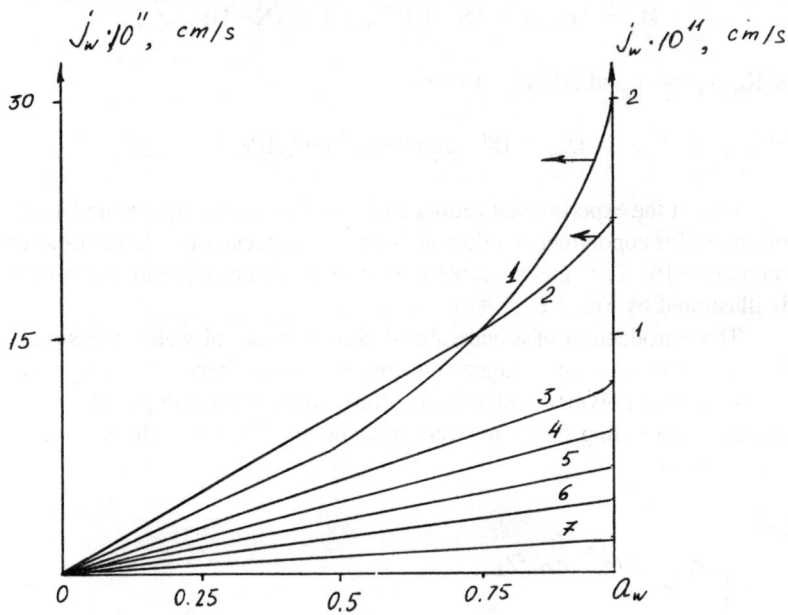

Figure 1.5. Dependence of water permeability through polymer films on water activity in the vapor phase. Fluoroplastics = F-2M (1), F-32L (2), F-10 (3), F-4 (4), F-4MB (5).

Eq. 1.9 allows us to write the expression for the integral diffusion coefficient of water

$$D_w = D_{wf} [(1 + K_T a_w)/(1 + K_{HT} a_w)]^{N-1} \times$$
$$(1 + (N+1)K_T a_w)/(1 + (N+1)K_{HT} a_w). \quad (1.10)$$

The partial diffusion coefficient of every associate of i molecules is equal

$$D_{wi} = D^0_{wf} \exp (-iV_w/BV_f)$$

with particular case of diffusion coefficient of free water

$$D_{wf} = D^0_{wf} \exp (-4\pi R^3_w/BV_f).$$

Eq. 1.10 may be simplified to

$$D_w = D_{wf} (1 + (N+1)K_T a_w)/(1 + (N+1)K_{HT} a_w)$$

if $K_{HT} a_w \ll 1$ and $NK_T a_w \sim 1$ or to

$$D_{wi} = D^0_{wf} \exp(-(n+1)V_w/BV_f).$$

One of the experimental results which confirmed the ligand model was observed for copolymer of ethylene with vinyl acetate of different ratio of monomers [6]. The agreement between theoretical and experimental results is illustrated by Fig. 1.6.

The introduction of a generalized cluster model of water diffusion is based on the reasonable suggestion that interaction between water molecules exceeds polymer–water interactions. Inculcation groups can be the second contributing factor in water association. Therefore, the most general cases are:

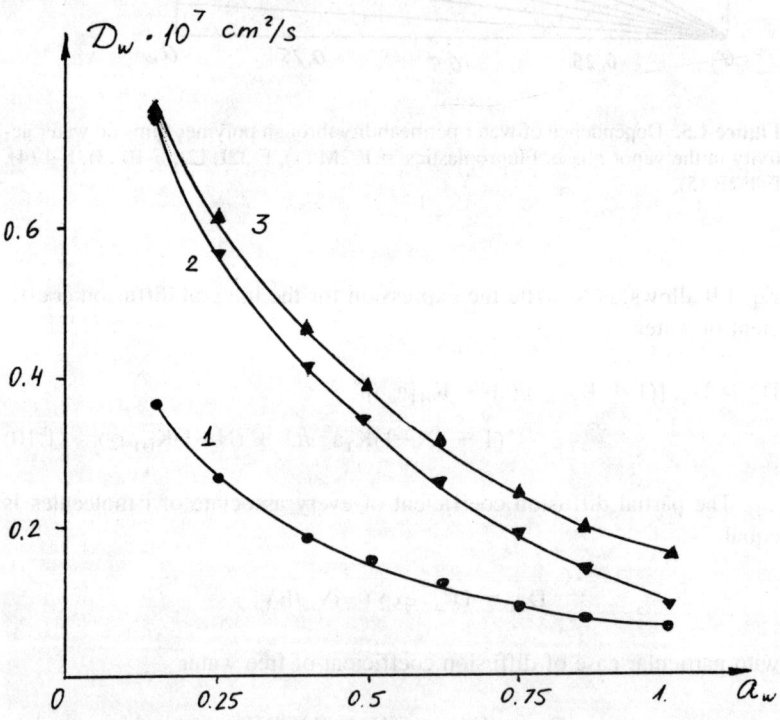

Figure 1.6. Diffusion coefficient of water in PE/VA copolymers of molar ratio: 85;15 (1), 75:25 (2), and in PVC (3).

—clusters are not formed, and water molecules are uniformly distributed in noncrystalline hydrophobic matrix;

—clusters are formed due to water–water interaction;

—inculcation groups contribute in cluster formation.

The above model allows us to explain many experimental results and
apply fundamental investigation in practice.

Water vapor was used as a diffusional probe to study segmental
mobility in hydrogen-bonded polymer blends. Modified polystyrenes containing 4-hydroxystyrene as comonomer units were blended with poly-
(methyl methacrylate). The diffusion coefficient of water decreased with
increasing concentration of the diffusant. Clustering of water molecules in
the polymer rendered a fraction of the sorbed water comparatively immobile. A partial immobilization model was used to estimate the diffusion
coefficient of the clusters. Interpolymer hydrogen bonding acted as physical crosslinks, which decreased the diffusivity of clusters. However, the
positive excess volume of mixing for one blend composition seemed to
have nullified the restrictive effect of hydrogen bonding on cluster diffusivity [34].

Moisture diffusion in unsaturated polyester, vinyl ester, and acrylic
resins exposed to various relative humidity and temperatures was investigated. The effects on water uptake of the type and amount of initiator, the
post-curing treatment, and specimen thickness were evaluated. A weight
loss during exposure was observed in some resins; the true equilibrium water content and the diffusion coefficient could not be determined for these
polymers. Absorption-desorption-reabsorption (ADR) experiments were
conducted to determine the true equilibrium water contents and diffusion
coefficients. These ADR experiments also showed whether a resin exhibits
a continuous weight loss during long-term water exposure, a phenomenon
associated with degradation of the polymer network, or whether the loss is
caused only by the release of residual species. The water diffusion behavior for polymers containing disc cracks was compared with the behavior
of polymers without disc cracks. Diffusion coefficients and saturation values were given for orthophthalic polyesters, isophthalic polyesters, vinyl
ester, and acrylic resins at 65°C and a 75% relative humidity [35].

A filled epoxy resin used as a structural adhesive and based on the
diglycidyl ether of bisphenol A cured with dicyandiamide was subjected in
its bulk form to aging at 40, 55, and 70°C and 100% relative humidity.
Gravimetric, viscoelastometric, and Fourier transformation infrared (FTIR)
studies have been affected after various times of exposure. Water absorption in the polymer was essentially Fickian, although closer inspection revealed the finer behavior to be sigmodal. The activation energy for diffusion was of the order of 80 kJ mol^{-1}, but there appeared to be no clear

relationship between equilibrium absorption values and temperature. Viscoelastometry showed that the glass temperature T_g diminishes from 115°C before aging to 90°C at saturation, 1% of water uptake corresponding to 8°C reduction in T_g. Reductions in Young's module were observed both in the glassy and rubbery states after aging and the latter is associated with molecular chain scission. FTIR analysis showed several modifications occurring due to water absorption, the main one being an increase in intensity of the band at 1740 cm-1. It was concluded that water absorption led both to plasticization effects and chemical modification of the epoxy resin [36].

Some difficulties may appear under study of composites and polymer blends. The water absorption and desorption behavior of poly(isobornyl methacrylate) and poly(tetrahydropyran-2-ylmethyl methacrylate) obeyed diffusion laws on repeated absorption/desorption cycles. However, the polymers of 2,3-epoxypropyl, tetrahydrofurfuryl, and tetrahydropyranyl methacrylates did not obey diffusion laws, did not equilibrate after 2 years of immersion in water, and exhibited very high uptake values (30–90%). For 2,3-epoxypropyl methacrylate, the sample disintegrated. A clearly detailed structure of the heterocyclic ring is critical. The use of these monomers in room temperature polymerizing poly(ethyl methacrylate)/monomer systems generally reflected the behavior of the related homopolymers [37].

Sorption of water vapor into thin films of a crosslinked dimethacrylate UV-cured polymer was measured by means of a quartz crystal microbalance apparatus. The diffusion coefficient and solubility were obtained. An upper limit to the water vapor diffusion coefficient, namely, the diffusion coefficient for liquid water in intimate contact with the sample, was measured for reference purposes by both gravimetric and infrared spectral techniques. A water uptake value for thick films was obtained gravimetrically as a reference for the thin film value. Measurement of the diffusion coefficient allows the determination of the length of time that the photopolymer could act as an effective water vapor barrier [38].

1.2.2. Water diffusion in hydrophilic polymers

To begin consideration of the fundamental aspects of water transport in hydrophilic polymers, we would like to start with a quotation from one of the most famous contributors in this area, N.A. Peppas. "Understanding the mechanisms of water sorption and diffusion in amorphous polymers is of particular importance in relation to our ability to interpret related transport phenomena in foods. Penetrant transport in rubbery or glassy polymers may be classified as Fickian, non-Fickian (anomalous), Case II or super Case II transport, depending on the thermodynamic activity and temperature of the system in consideration. Additional structural parameters influencing the transport mechanism include the molecular weight, degree of crosslinking and degree of branching of the polymer, and its thermal and solvent expansion coefficients. The mechanisms of water transport can be determined by a variety of experimental techniques, the simplest one of which is gravimetry. In gravimetric studies, the amount of water absorbed in the polymer is plotted versus time according to equation $M_t/M_{oo} = kt^n$. If the exponent n is 0.5 (for planar systems) the diffusion is Fickian. Non-Fickian behavior is observed for $0.5 < n < 1.0$, with a limit of Case II transport for $n = 1.0$. Non-Fickian and Case II transport are indicative of coupling of diffusional and relaxational mechanisms. Relaxation is related to a transition from rubbery to glassy state. Major relaxational mechanisms are indicative of stresses formed in the polymer during swelling. These observations translate into important consequences for water transport in food products, as the mechanism of water transport may influence the performance of food products in processing and use" [39].

A system of partial differential equations was developed to describe the transient, three-step transport of moisture through packaged food products. The steps included (i) Fickian diffusion through the polymer package film; (ii) Langmuirian adsorption upon the food surface; and (iii) Fickian diffusion through the food material. A set of finite difference equations was derived to approximate the continuous model. These equations were solved for standard boundary conditions in each section of the food packages. The results can be used for the determination of the food package shelf life [40].

Water sorption which is not classically Fickian was observed in a variety of polymers. Deviation from Fickian kinetics was widely assumed to be caused by rate-limiting polymer relaxation, despite minimal proof of this. To the contrary, the evidence accumulated in [13] indicated that water transport in initially glassy poly(2-hydroxyethyl methacrylate) (PHEMA), an important water-swellable biomedical polymer, is controlled by Fickian diffusion. First of all, the fractional water uptake was initially linear and independent on sample thickness when plotted against the square root of time over initial thickness, as expected for a Fickian process. Furthermore, the moving solvent front also advanced with the square root of time. Temperature, polymer thermal history, and initial solvent concentration all affected the sorption kinetics of PHEMA in manners consistent with a Fickian process. The invariably Fickian sorption mechanism was believed to be the consequence of the water molecule's small size and affinity for hydrophilic, swellable polymers [41].

The kinetic regularities of water sorption in hydrophilic polymers are the same as in other solvent-polymer systems. Therefore, the same models can be applied.

Wu and Peppas [42] developed the mathematical model to explain the anomalous penetrant diffusion behavior in glassy polymers. The model equations were derived by using the linear irreversible thermodynamics theory and the kinematic relations in continuum mechanics, showing the coupling between the polymer mechanical behavior and penetrant transport. This phenomenon is particularly important in glassy polymers. The Maxwell model was used as the stress-strain constitutive equation, from which the polymer relaxation time was defined. An integral sorption Deborah number

$$De = \tau/\nu_D$$

was proposed as the ratio of the characteristic relaxation time (τ_D) in the glassy region to the characteristic diffusion time in the swollen region (ν_D). With this definition, an integral sorption process was characterized by a single Deborah number, and the controlling mechanism was identified in terms of the value of the Deborah number. The model equations were two coupled nonlinear differential equations. A finite difference method was developed for solving the model equations. Numerical simulation of integral sorption of penetrants in glassy polymers was performed. The simulation results showed that (i) the present model can predict Case II transport behavior as well as the transition from Case II to Fickian diffusion, and (ii) the integral sorption Deborah number is a major parameter affecting the transition. In [43] Wu and Peppas introduced a new numerical algorithm that might be used to analyze complex problems of penetrant transport. They assumed that penetrant transport in polymers often deviates from the predictions of Fick's law because of the coupling between penetrant diffusion and the polymer mechanical behavior. The model consisted of two coupled differential equations for penetrant diffusion and polymer stress relaxation, respectively. If the polymer relaxation is the rate-limiting step, both the concentration and stress profiles are very steep. An algorithm based on a finite difference method was proposed to solve the model equations. It features the development of a tridiagonal iterative method to solve the nonlinear finite difference equations obtained from the finite difference approximation of the differential equations. This method was found to be efficient and accurate. Numerical simulation of penetrant diffusion in glassy polymers was performed, showing that the integral sorption Deborah number is a major parameter affecting the transition from Fickian to anomalous diffusion behavior.

Petropoulos proposed [44] and then developed [27, 45] the differential swelling stress model which describes the transport of solvent by the diffusion equation of standard form

$$\delta C_w/\delta t = (\delta/\delta x)(D_w \delta C_w/\delta x) \qquad 0 < x < 1 \qquad (1.11)$$

where

$$D_w = D_w \exp(k_{w1} C_w + k_{w2} f) \qquad (1.12)$$

coupled with an equation describing the build-up and relaxation of the corresponding local differential swelling stresses (f) along the plane of the film (i.e. at right angles to the direction of solvent penetration)

$$\delta f / \delta t = (G_o - G_{oo}) \delta s / \delta t + \delta (s G_{oo}) / \delta t +$$

$$((G_o - G_{oo})^{-1} \delta (G_o - G_{oo}) / \delta t - \beta)(f - s G_{oo}). \qquad (1.13)$$

The change in area (A) caused by unconstrained solvent uptake is given by

$$A(C_w) = A(0)(1 + k_s C_w). \qquad (1.14)$$

During the sorption of solvent, its concentration (C_w) varies with x and a corresponding variation of A according to Eq. (1.14) is expected. However, the actual area of a thin film is constrained to a uniform value $A(t)$, which is a function of time only. This leads to the creation for local strains

$$s = A(t)/A(x,t) - 1$$

and corresponding local stresses f, which must add up to zero net overall stress. Hence, we have

$$\int_o^l f(x,t)dx = 0.$$

The plasticizing action of solvent is described by exponential dependence of the instantaneous (G_0) and long term (G_{oo}) elastic modules, and relaxation frequency (β) on solvent concentrations

$$G_o = G_{oo} \exp(-k_g C_w)$$

$$G_{oo} = G_{ooo} \exp(-k_g C_w)$$

$$\beta = \beta_o \exp(k_{d1} C_w + k_{d2} C_s).$$

Some of the main results obtained by Petropoulos are shown in Figs. 1.7 and 1.8. In both cases, line 1 is "Fickian" and the deviation from this curve increases as the effect of stress becomes more marked. The pattern followed is much the same whether the increasing effect of stress is manifested by (i) increase of the stress set up or the stress-dependence of D_w (curve 1 - curve 7 - curve 4), (ii) a reduction in the extent of plastization

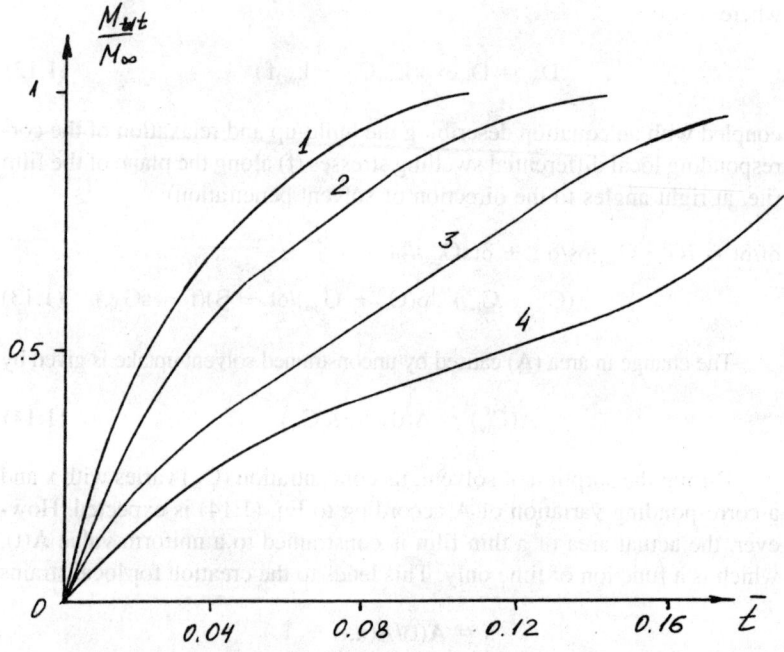

Figure 1.7. Computed absorption kinetic curves. Effect of softening. Curve 1 is Fickian. $G_o(1)/G_o(0) = 0.1$ (2), 0.4 (3), 1 (4).

(or "softening") of the polymer by the penetrant (curves 1-4 in Fig. 1.7), or (iii) a decrease in the rate of stress relaxation as compared with the rate of diffusion (curve 1 - curves 5-7 in Fig. 1.8). In each case, as the deviation from "Fickian" behavior becomes more marked, the M_t/M_{oo} vs t curve acquires a quasi-linear middle portion which first extends gradually and finally becomes concave upwards. The general effect was one of reduction of the overall rate of sorption and is greatest in the case of curve 4, which is characterized by the absence of softening, practically no stress relaxation, and the strongest dependence of D on f. The well-defined "Super Case II" character of this curve is due to strong acceleration of the diffusion process in the later stages of sorption.

Further development of this model [46] showed that the acceleration of water sorption coupled with minor changes of the shape of kinetic curve is the main result of the strong dependence of D_w on C_w (Fig. 1.9). Because C_w and f effect D_w in opposite directions their effects tend to cancel at low M_{wt} (overall water uptake) values. However, there is substantial dif-

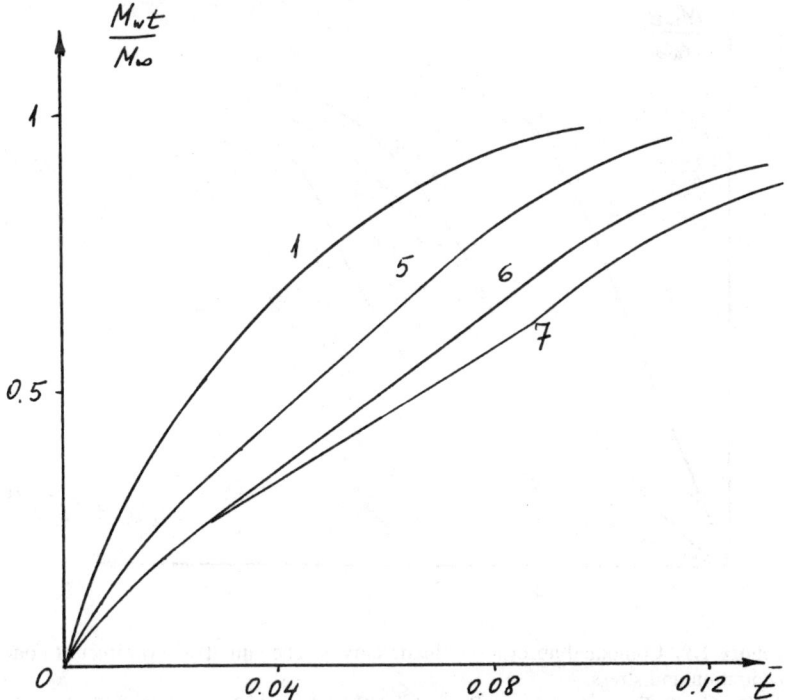

Figure 1.8. Computed absorption kinetic curves. Effect of relative stress relaxation frequency. Curve 1 is Fickian. $\beta(0) = 10$ (5), 1 (6), 0.1 (7).

ference between the cases of strong dependencies of D_w on C_w and f at high M_{wt} values. Acceleration of water sorption is the consequence of the increase of b_o, and the shape of the curve becomes Fickian (at the same time the profile of water in the film remains sharp because of strong dependence of D_w on C_w).

The results of theoretical study are relevant to many experiments. The main known kinetics curves can be reproduced satisfactory under well-defined conditions as we did it, for example, under study of water sorption in PVA [47], cellulose derivatives [48], copolymer of N-vynilpirrolidone and butyl (or methyl) methacrylate [47, 49], and copolymer of maleic anhydride with vinyl acetate [24].

Poly-ether-ether-ketone (PEEK) is a thermoplastic tough polymer used as a matrix for advanced composite materials in aeronautic applications. The investigation of its resistance to humid environment and to exposure

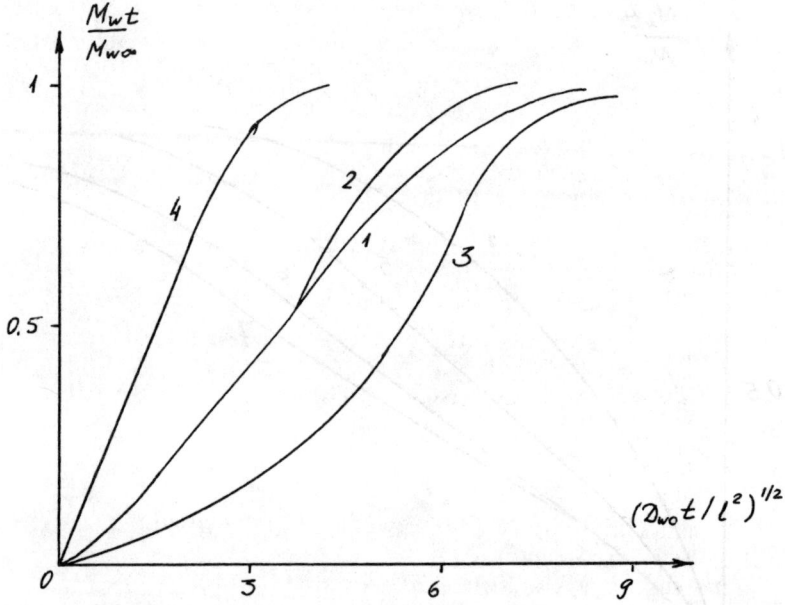

Figure 1.9. Computed absorption kinetic curves. The simultaneous effect of concentration and stress.

1. $D_w(C_{wo},0)/D_{wo} = e$; $D_w(0, f_{max})/D_{wo} = \exp(f_{max})$; $\beta_o = 1$
2. $D_w(C_{wo},0)/D_{wo} = 20000$; $D_w(0, f_{max})/D_{wo} = \exp(10f_{max})$; $\beta_o = 1$
3. $D_w(C_{wo},0)/D_{wo} = e$; $D_w(0, f_{max})/D_{wo} = \exp(10f_{max})$; $\beta_o = 1$
4. $D_w(C_{wo},0)/D_{wo} = e$; $D_w(0, f_{max})/D_{wo} = \exp(10f_{max})$; $\beta_o = 100$

to organic and chlorinated solvents is extremely important. A series of integral sorption experiments were performed on PEEK with three different kinds of penetrant, namely, water, methylene chloride, and methylene chloride-n-heptane mixtures. Coupled diffusion and relaxation phenomena were observed to occur in most of the adopted experimental conditions. The sorption behavior was found to range from ideal Fickian diffusion (where the molecular mobility is not affected by the penetrant concentration) to so-called anomalous diffusion. In the latter case the penetrant mobility was a complex function of penetrant concentration, time, and temperature. An overview of theoretical models reported in the literature to describe such phenomena was also presented with the aim of interpreting the experimental results obtained [50].

The application of free-volume concepts to describe diffusion in polymer-solvent systems was reviewed recently by Duda [51]. He showed how the Vrentas-Duda free-volume model for diffusion in polymer solutions above the glass transition temperature can be modified to describe diffusion of low molecular weight species in glassy polymers. It was discussed how free-volume concepts can be extended to describe antiplasticization behavior which occurs when molecular motions in glassy polymers are retarded by the addition of certain low molecular weight species.

The Vrentas/Duda free-volume diffusion model accurately correlates polymer/solvent diffusion coefficients over wide ranges of concentration and temperature. Currently the model is semipredictive: limited diffusion data are required to estimate model parameters that can then be used to predict diffusion coefficient behavior over sundry conditions. In [52] Zielinski and Duda presented methods for estimating all of the model parameters without any diffusion data and examine the accuracy of the resulting diffusion coefficient predictions. This is the only technique known that predicts polymer/solvent diffusion behavior without use of any diffusion data.

Lee and Kim [53] resolved the fundamental issue whether a penetrating solvent front would maintain a constant rate throughout a radially symmetric polymer sample, such as a sphere or a cylinder, under conditions where the front penetrating behavior in a sheet sample of identical polymer composition and comparable dimension was completely Case II. Similar concerns on the effect of geometry in systems with known Fickian and non-Fickian sorption behavior were also addressed there. These issues were dealt with by establishing a theoretical framework with relevant front penetration behavior simulated for different geometry, based on available solutions to the corresponding moving boundary problem. This was followed by presenting experimental results comparing swelling front movement in sheet and spherical bead samples of identical polymer composition and morphology. In this regard, their experimental data on water penetration in poly(2-hydroxyethyl methacrylate) (PHEMA) and methanol penetration in poly(methyl methacrylate) (PMMA) showed very good agreement with the predicted trends. Both the experimental and theoretical results presented there revealed that only the initial stage of the solvent penetration in radially symmetric geometry exhibits same transport characteristics as in sheet samples. This was followed by an intermediate transition region with an apparent linear front movement prior to an accelerated front penetration towards the core. In addition to demonstrating that such front acceleration is a natural outcome of the radially symmetric geometry, the results also suggest that it occurred independent of the transport mechanisms involved.

A mathematical model was presented in [54] to describe non-Fickian diffusion with chemical reaction of a penetrant A with some reactive group B of a granular glassy polymer. The diffusion process cannot be described by Fick's law due to the swelling of the glassy polymer grain caused by the penetration of the diffusing species. Therefore, the kinetics of swelling must be taken into account. A new model was presented, allowing for mass transfer with chemical reaction while assuming power-law kinetics for the velocity of the swelling front:

$$dr_0/dt = -K(c(a,r = r_0) - c(a)^*)n$$

where $c(a)^*$ is the threshold concentration for swelling, $c(a,r = r_0)$ the concentration of penetrant A at the position r_0 of the swelling front, and K a constant independent of the penetrant concentration. Two cases of reaction kinetics were analyzed: first order in A only and first order in both A and B (overall second order). Criteria for the occurrence of homogeneous addition and for a shrinking core type of reaction are given. For a low conversion of B, the reaction can be pseudo-first-order in A. Both Fickian diffusion (high K, n_{00} and $c(a)^* = 0$) and the so-called case II diffusion (low K, n_0) were shown to be asymptotic solutions of the model. It was also shown that in some cases an excellent agreement can be obtained between the numerical unsteady-state solution and the so-called pseudo-steady-state solution in which the accumulation term of the mass balance of A is neglected. This PSS-solution could be obtained analytically. Both the case without reaction and with first-order kinetics in A are analyzed. The approximation is rather good for high first-order reaction rates, short diffusion times, and if case II diffusion was approached [54].

Best described with coauthors the design and implementation of a method based on infrared monitoring of the OD stretch mode which permits measurement of the diffusion profiles of water permeating glassy polymers for the first time. Results for several glassy polymers were compared with a theory of trapping diffusion which leads to a nonlinear diffusion problem in which the trapped water penetrates the polymer as a sharp front if there is a large amount of trapping. In the case of two of the polymers studied, the trapping model describes the data, very well (much better than a simple diffusion model) and permits determination of parameters related to the free volume and the residence time of the water on traps in the model [55].

Certain difficulties arise with a description of the diffusion of water in so-called moderately hydrophilic (or moderately hydrophobic) polymers (the title depends on the desire or habit of authors). These polymers take from 1 to 10% of water and can show features either of hydrophilic or hydrophobic polymers.

The electrochemical impedance spectroscopy (EIS) technique was used to evaluate the water transport (diffusion and equilibrium water uptake) and the dielectric properties of freestanding polyimide (Kapton(R) and PMDA-ODA) and polyethyleneterephthalate (PET) membranes at 25 and 40°C, respectively, and in supported pyromellitic dianhydride-4,4'-oxydianiline (PMDA-ODA)-coated metals. Permeability and diffusion coefficients of freestanding films were also obtained by using the Payne cup method and the MacBain quartz spring balance in order to assess the reliability of the EIS method when compared to other techniques. Results from this work show that the diffusivity of water in polyimide films varies from 1.4×10^{-9} cm^2/s to 3.5×10^{-9} cm^2/s for thickness between 2.4 and 125 μ, while the equilibrium water uptake varies from 2.31 to 4.63% by weight for the same range of thickness. The average calculated dielectric constant of the freestanding Kapton(R) films is 4.5. Water diffusion coefficient in PET varies from 2.1×10^{-9} cm^2/s to 12.6×10^{-9} cm^2/s for thickness between 22 and 205 μ, while the equilibrium water uptake varies from 0.50 to 0.95% by weight for the same range of thickness. The average calculated dielectric constant of the freestanding PET films is 3.6. Transport properties' results obtained through the electrochemical technique are in reasonable agreement with those obtained with the classical gravimetric method. Capacitance data obtained with the EIS technique were interpreted in terms of a model in which the electrical analog of the film is composed of either a parallel RC or of a series of parallel RC circuits. Thick films are better described in terms of a series of parallel RC circuits. Theoretical analysis presented in this paper indicates that transport properties of water through freestanding and applied films (paints) can be obtained from EIS (capacitance data) only when specific requirements are met [56].

The aqueous kinetic swelling properties of a class of cross-linked hydrophobic polyamine copolymer gels based on n-alkyl esters of methacrylic acid (nAMA) and N,N-dimethylaminoethyl methacrylate (DMA) were studied as a function of solution ionic composition (pH, ionic strength, and buffer content), gel composition, and temperature. Water uptake and swelling in these gels are driven by ionization of the DMA amine groups, which overcomes the hydrophobic tendency of these gels to exclude water in the unionized state. Sorption kinetics in initially glassy gel disks were generally biphasic, characterized by an initial phase of relatively slow water uptake followed by an accelerated phase during which significant volume expansion of the gel occurs. This sorption/swelling behavior strongly suggests a moving penetrant front mechanism. The initial rate of water sorption increases markedly as (i) solution pH decreases, (ii) gel nAMA comonomer content decreases, (iii) gel nAMA side-chain length decreases, and (iv) temperature increases. Furthermore, the initial phase of sorption in initially glassy gels is generally non-Fickian and approaches zero-order behavior as (i) pH increases, (ii) nAMA content increases, and (iii) temperature decreases. In direct contrast, sorption in initially dry, rubbery gels is monophasic, but non-Fickian, and approaches zero-order behavior as temperature increases. This behavior is contrary to the Fickian sorption behavior normally observed in polymers above their glass transition temperatures. Finally, sorption kinetics critically depended upon the nature of the ions in solution. Kinetics were significantly faster in the presence of weak electrolytes than that of strong electrolytes [1].

A bending-beam technique was used to in situ monitor the diffusion of water in various polyimide films. The polyimides studied were PMDA-ODA, pyromellitic dianhydride-p-phenylenediamine (PMDA-PDA), and 3,3′,4,4′-benzophenone tetracarboxylic dianhydride-p-phenylenediamine (BPDA-PDA), and their blends and random copolymers. The diffusion of water in these films obeys Fick's law. In PMDA-ODA, the effective diffusion coefficient was found 5×10^{-9} cm^2/s for thickness ranging from 6.7 to 27.3 μ. In PMDA-PDA, it is 2×10^{-9} cm^2/s for thickness ranging from 7.3 to 20.0 μ, and in BPDA-PDA, 0.2×10^{-9} cm^2/s for thickness ranging from 4.8 to 21.0 μ. In the blends and random copolymer with 50% wt PMDA-ODA and 50% wt PMDA-PDA, the diffusion constants were slightly smaller than those in the pure PMDA-ODA, but much larger than in the pure PMDA-PDA. On the contrary, in those with 50% wt BPDA-PDA and 50% wt PMDA-PDA, the diffusion constants were much smaller than those in the pure PMDA-PDA, but slightly larger than in the pure BPDA-PDA. These diffusion constants were primarily affected by the chemical structure of the imide molecule. The morphology, such as crystallinity, of the films has played a secondary factor. Hygroscopic stresses due to water uptake in all the studied films increase as the film thickness increases. It can be attributed to the fact that the film orientation decreases with the increase of thickness [57].

The uptake of water by nylon 6,6 [42DB Adipure (trade name of Dupont Canada Inc.)] at 100°C has been monitored by a combination of one-dimensional proton NMR spectroscopy, relaxation time (T_1 and T_2) measurements and proton microscopic NMR imaging techniques. The relaxation times of the water absorbed into the nylon matrix are very short at room temperature, ($T_2 < 1$ ms and T_1 almost equal to 1 s) indicating that the water is located in a highly restricted environment and suggesting that strong interactions exist between the absorbed water and the polymer. The diffusion profiles measured at room temperature indicate that the diffusion of water into nylon 6,6 at 100°C is Case I Fickian diffusion. The spatial dependence of the T_2 relaxation time constant and its variation with the water content was also examined. The results reveal that both T_2 and T_2^* decrease toward the center of the sample in samples that have a concentration gradient of sorbed water. In fully saturated samples, no spatial dependence was observed. The overall values of T_2 and T_2^* are also observed to increase as a function of exposure time. An evaluation of the desorption process at room temperature and at 100°C was performed. A continuous, exponentially decreasing solvent profile was observed for the desorption process which again indicates Case I Fickian kinetics. The exchange process of external bulk and atmospheric water with deuterium oxide (D_2O) saturated nylon rods has also been studied using the microscopic imaging technique [58].

The density and dimensional changes of Nafion 117 H were measured as a function of the water content. Swelling of Nafion commences at N (= mol ratio of water to hydrogen ion) was found equal approximately to 1.9. Some anisotropy of dimensional changes was observed. Water-sorption isotherms obey Henry's Law with a nonzero intercept indicating some water retention. The diffusion coefficient of water in Nafion and the electrical conductivity of Nafion were strong functions of the water content. The latter is exploited for the development of a humidity sensor [59].

A model edible film made with methylcellulose was studied for its water vapor barrier properties. The steady-state water vapor transmission rate (WVTR) increased with both the water vapor pressure gradient and the initial water content before permeation. A decrease in water diffusion with increasing moisture content was due to a clustering phenomenon of water molecules within the film. Water concentration profiles within the film were estimated from the sorption isotherm, and differed from the theoretical linear profile based on Fick's first law. The WVTR and the diffusivity depended strongly on the water concentration because of interactions between water molecules and the polymer matrix. Thus water vapor permeability (WVP) calculations used for synthetic polymeric packaging and based on Fick's and Henry's laws did not apply for methylcellulose edible films [60].

The problem of describing quantitatively the process of diffusion in polymers attended by plasticization of their amorphous phase was considered in [61]. A consequence of such plasticization was that the volume accessible to the penetrant molecules continuously increased during the course of their sorption. Assuming that the rate of relaxation processes was far greater than that of diffusion, an equation had been generated to describe the sorption in a matrix with varying accessibility. A numerical solution of this equation, using experimental data for a polyamide-water system, demonstrated that sorption in a matrix with increasing accessibility was slower than sorption in a hypothetical matrix with constant accessibility and with the same proportion of the amorphous phase. At the same time, the concentration of the penetrant in the former case was higher at any point in the polymer specimen. The results obtained are important for calculating the rate constants for chemical reactions that proceed in a polymer matrix in the diffusion-kinetic mode [61].

The results of the interphase diffusion in the system hydroxyethylcellulose film–water as a function of solid polymer compression were discussed in [62]. The experimental technique was briefly described and the possible swelling mechanism was proposed. The pressure dependence of the diffusion coefficient was divided into two regions: one of elastic and the other of permanent polymer chain deformation. The change to permanent deformation was controlled by deformation time or by values of the external pressure.

The study of micromolecular transport in cellulose acetate film was extended in [63] by investigation of the effect of different degrees of semipermanent uniaxial macromolecular orientation, produced by prestretching the film under suitable conditions. The changes noted in penetration rate and kinetics and in birefringence, tracer microdensitometry, and microinterferometry profiles for penetration along and across the axis of preferred orientation were reported in detail and their implications in terms of the transport mechanism were discussed [63].

A new polymer dissolution model was developed by incorporating the polymer chain disentanglement mechanism into the relevant transport equations. The disentanglement time was used as a dissolution characteristic time controlling the moving position of the solvent-polymer boundary. A dimensionless dissolution number was defined as the ratio of the characteristic polymer disentanglement time to the characteristic solvent diffusion time. The dissolution number was shown to be proportional to the square of the gel layer thickness. Scaling law expressions for the dependence of the gel layer thickness and the polymer dissolution rate on polymer molecular weight were also derived. Solution of the model for one-dimensional dissolution showed three distinct dissolution stages and confirmed the proposed scaling law relations for the gel layer thickness and the dissolution rate.

Experimental studies of dissolution of polystyrene and poly(methyl methacrylate) (PMMA) in methyl ethyl ketone (MEK) were used to verify the model, and two types of polymer dissolution behavior were observed. For dissolution of polystyrene in MEK, the solvent diffusion behavior was Fickian and a constant gel layer thickness was observed during the stationary dissolution stage. The effect of polymer molecular weight on the gel layer thickness was investigated for nine monodispersed samples, with MBAR(n) ranging from 28,000 to 2,830,000. The experimental results showed that the dependence of the gel layer thickness on molecular weight is more prominent in the high molecular weight region. The polystyrene data verified the new dissolution model. The dissolution of PMMA in MEK was controlled by crack propagation as no significant gel layer was observed [64].

Mathematical models were developed to describe moisture transport in epoxy composites using concentration and stress dependent diffusion coefficients. These models take into consideration the stresses induced due to moisture and their influence on the transport mechanism. All governing differential equations are solved with appropriate boundary conditions to obtain concentration and stress profiles. It is concluded that non-Fickian and anomalous transport of moisture in epoxy composites can occur under certain conditions, leading to stress accumulation close to the fibers of the composite [65].

The linear (unusual) character of the dependence of diffusion coefficient of water in some cellulose derivatives (Fig. 1.10) was explained in [48] by their intermediate properties and by two opposite effects, which are swelling of polymer matrix causing the acceleration of transport of water and formation of clusters slowing it down.

The polymer matrix, divided in a number of cells in which the penetrant molecules can be sorbed and migrate, was considered in [66]. Each cell was assigned an effective energy value that obeys a particular distribution. The effective diffusion coefficient and its concentration and temperature dependence were determined. The origin of sorbed penetrant mobility was also investigated. Using a delta-Dirac distribution for the site's

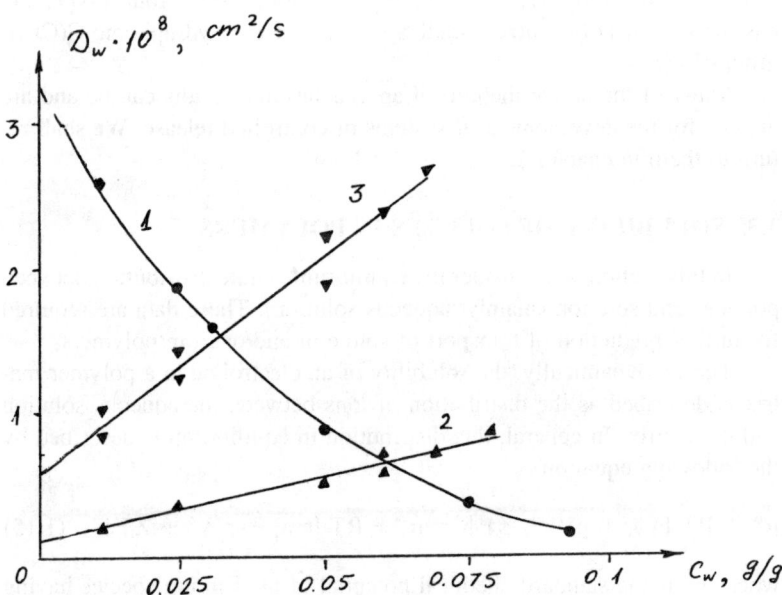

Figure 1.10. Concentration dependence of water diffusion coefficient in cellulose derivatives: cellulose diacetate (1), cellulose acetyl fluorate, CAF (2), CAF blend with cellulose triacetate (3).

energetic values, the model is reduced in the appropriate limit (low pressure) to other formulations of the dual transport model. More general results, allowing the site's energetic values to be drawn from a Gaussian distribution, were also given in [66].

Sorption and diffusion of water vapor were investigated gravimetrically for polyimide films. The activity dependence of the solubility and diffusion coefficients, S and D, respectively, was classified under four types: (1) constant S and D type, (2) dual-mode sorption and transport type, (3) dual-mode type followed by a deviation due to a plasticization effect at high vapor activity, and (4) constant S and D type followed by a deviation due to water cluster formation at high activity. For the dual-mode type, the Henry's law component is much larger than the Langmuir component except at low activity, and, therefore, deviation in behavior from the first type is small. S is larger for polyimides with higher content of polar groups such as carbonyl, carboxyl, and sulfonyl. D is larger for polyimides with a higher fraction of free space, with some exceptions. The polyimide from 3,3',4,4'-biphenyltetracarboxylic dianhydride and dimethyl-3,7-diaminodibenzothiophene-5,5-dioxide belongs to the third type and displays both large S and large D. The polyimide from 2,2-bis (3,4-dicarboxyphenyl) hexafluoropropane dianhydride and 4,4'-oxydianiline belongs to the fourth type, and has the largest D but rather small S because of the hydrophobic $C(CF_3)_2$ groups [67].

Many of the above theoretical approaches and results can be and are applied for the development of systems of controlled release. We shall return to them in chapter 2.

1.3. SOLUBILITY OF SOLUTES IN POLYMERS

In this section, we consider the equilibrium solute distribution between polymer and solution (mainly, aqueous solution). These data are required for further prediction of transport of solute in and/or from polymers.

Thermodynamically, the solubility of an electrolyte in a polymer matrix is described as the distribution of ions between the aqueous solution and the matrix. In general, this distribution in equilibrium is described by the following equation

$$\mu_i^o + RT \ln a_i + pV_i + Z_iF\phi = \mu_i^{o'} + RT \ln a_i' + p'V_i' + Z_iF\phi' \quad (1.15)$$

where μ_i^o is the standard chemical potential of the i-th ion species having charge Z_i, V_i is the partial molar volume of the ion, and p and ϕ are the pressure and electric potential, respectively [6]. The right side of the equation relates to the polymer phase. Usually, to simplify the treatment, V_i is assumed to be the same in both phases, and ion concentration in polymer C_i is used instead of its activity a_i.

The thermodynamic solubility coefficient is introduced as $S_i = C_i/a_i$, that is different from ion distribution coefficient,

$$K_{pi} = C_i/c_i = K_i \psi f_\pi \tag{1.16}$$

or the chemical coefficient of distribution $K_i = \gamma_i / \gamma'_i \exp [(\mu_i^{o'} - \mu_i^o)/RT]$.

Other parameters involved in expression 1.15 are the following:

$\psi = \exp [ZF(\phi' - \phi)]$ is the electrostatic factor of the distribution coefficient;

$f_\pi = \exp [V_i (p' - p)]$ is the contribution of the swelling pressure to the distribution coefficient;

γ_i is the activity coefficient; c_i is the concentration of solute in aqueous solution.

In the neutral medium, the distribution of solute $K_m A_n$ between polymer and water normally follows the Nernst equation

$$a'_s = K_p a_s$$

where

$$a_s = a_K^m a_A^n = C_C C_C g^{m+n}_\pm$$
$$K_p = (K_K^m K_A^n)^{1/(m+n)}.$$

Particularly for $m = n = 1$

$$K_p = (K_K K_A)^{1/2}.$$

The uranium sorption of an ascorbic chitosan derivative polymer (N.D.T.C. N-[2-(1,2-dihydroxyethyl) tetra-hydrofuryl] chitosan) was shown to be three to four times greater than chitosan under the same conditions, reaching 600 mgU/g (i.e., 2.5 mmolU/g-1 polymer). The effect of variation of the classic parameters was studied: pH, metal concentration, and particle size. pH and metal concentration, which induced a change in metal solution chemistry, were found to be the controlling parameters, mainly affecting intraparticle diffusion; the diffusion rate was strongly affected by ionic size of solute, as opposed to a limited effect on external mass transfer. pH plays an important part in equilibrium studies, and sorption isotherms were mainly a function of pH. The Freundlich model showed a better correlation coefficient than the Langmuir model when fitted to experimental results; the uptake mechanism involved a monolayer sorption with molecular interactions between molecules sorbed on surface [68].

The more complicated Donnan equation describes the electrolyte distribution between the solution and the polymer matrix, containing charged groups (proteins, polyamides, etc.).

The condition of electroneutrality in polymer containing ionogenic groups is written in form

$$\sum_{}^{p} Z_{Rl}C_{Rl} = \sum_{}^{m} Z_{Yi}C_{Yi} - \sum_{}^{n} Z_{Xj}C_{Xj}$$

where R, Y, and X correspond to the functional groups in polymer, counter-ion, and co-ion, respectively, and l, i, j generalize the possibility of existence of several types of ions and ionogenic groups.

For a case of equilibrium in the system of polymer–electrolyte the equality of electrochemical potential [69] or for pair chemical potential [70] may be written. Further simplification ($Z_X = Z_Y$) leads to the formula of Donnan equilibrium

$$C_Y = Z_R C_R / 2 Z_Y + [(Z_R/Z_Y)^2 (C_R/2)^2 + K_{Y,X} C^2_Y]^{1/2}$$

$$C_X = -Z_R C_R / 2 Z_X + [(Z_R/Z_X)^2 (C_R/2)^2 + K_{Y,X} C^2_X]^{1/2}$$

where the exchange coefficient is defined as

$$K_{Y,X} = (\gamma_\pm / \gamma'_\pm)$$

If concentration of solute in polymer (C_s) is relatively low ($C_s \ll Z_R$) we get

$$C_X \ll Z_R C_R, \; C_Y \ll Z_R C_R,$$

and finally

$$C_Y = Z_R C_R / Z_Y, \; C_X = Z^2_{RX} C^2_X K^{1/2}_{Y,X} / Z_R C_R.$$

For higher concentration of solute ($C_s \gg Z_R$, $C_X \gg Z_R C_R$, $C_Y \gg Z_R C_R$) we get the case which corresponds to the Nernst distribution

$$C_Y = K^{1/2}_{Y,X} C_Y \; C_X = K^{1/2}_{Y,X} C_X$$

where $K_{pX} = K_{pY} = K^{1/2}_{Y,X}$.

The theory of Donnan equilibrium was applied for the description of the sorption of the acid-salt mixture in hydrophilic polymer and for the creation of the relevant boundary condition for theoretical description of the transport of buffer mixture [71]. The sorption of two electrolytes with the same anion (KCl and HCl) in polyvinyl alcohol (PVA) was studied. The sorption isotherm of HCl had a linear part, positively diverged from line at $C_{HCl} > 0.3$ and rather weakly depended on KCl concentration (Fig. 1.11). In contrast, KCl sorption in PVA films depended on the equilibrium content of HCl (Fig. 1.12). The addition of acid led to the release of salt from PVA, and altered the shape of the sorption curve. This effect was explained in terms of Donnan equilibrium.

The protonization of hydroxyl groups in PVA was written in the following form

$$R—OH + H_3 + O \overset{K_F}{=\!=\!=} R—OH^{2+}_2 + H_2O$$

where the equilibrium constant (K_F) was defined as

$$K_F = C_{F+}C_w/C_FC_H = C_{F+}C_w/C_H(C^0_F - C_{F+})$$

and C_H was the concentration of free form of acids, C_F and C_{F+} were the concentrations of free and protonized groups of hydroxyls in PVA, re-

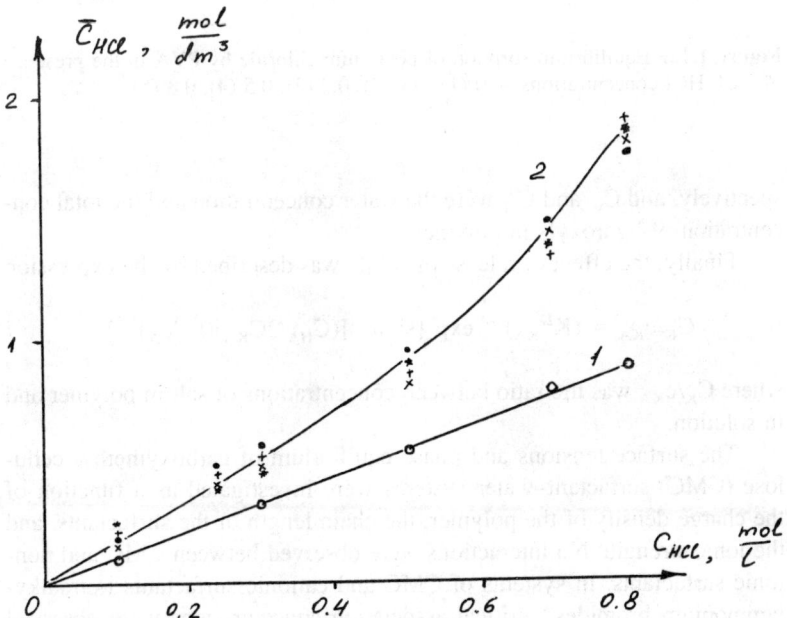

Figure 1.11. Equilibrium sorption of hydrochloric acid by PVA in the presence of KCl. KCl concentrations = 0 (1); 0.1, 0.3, 0.5, 0.8 (2).

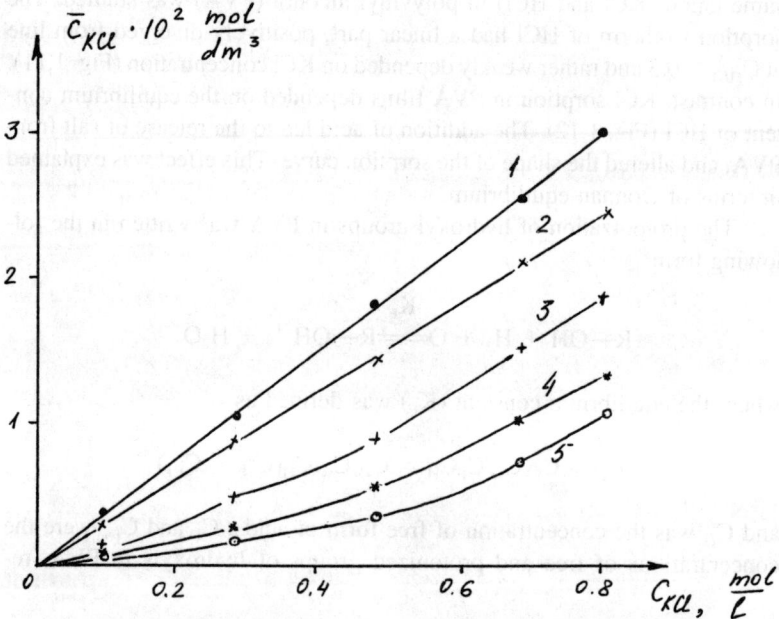

Figure 1.12. Equilibrium sorption of potassium chloride by PVA in the presence of HCl. HCl concentrations = 0 (1); 0.1 (2), 0.3 (3), 0.5 (4), 0.8 (5).

spectively, and C_w and C^0_F were the water concentration and the total concentration of hydroxyls in polymer.

Finally, the effect of release of solute was described by the expression

$$C_K/c_{KA} = (K^D_{KA})^{1/2} \exp \{\sinh^{-1}[(C_{HA}/2C_{KA})(K^D_{KA})^{1/2}]\}$$

where C_K/c_{KA} was the ratio between concentrations of salt in polymer and in solution.

The surface tensions and phase equilibrium of carboxymethyl cellulose (CMC)-surfactant–water systems were investigated as a function of the charge density of the polymer, the chain length of the surfactants, and the ionic strength. No interactions were observed between CMC and nonionic surfactants. In systems of CMC and cationic surfactants (tetraalkylammonium bromides), critical association concentrations were observed at concentrations well below the critical micelle concentrations of the surfactants. Associative phase separation occurred in extremely dilute systems when the charge ratio between the surfactants and the polymers was close to zero. The separated phase was a viscous gel phase containing 40–60% of water. The properties of these systems can be qualitatively understood assuming that (i) the driving force for association of the surfactants with the polysaccharide chains increases as the hydrophobic chain length of the surfactant or the charge density of the polyelectrolyte increases, and (ii) there is an associative interaction between the surfactant aggregates and the polysaccharide which decreases with increasing ionic strength [72].

Interactions between poly(acrylic acid) (PAA) and sodium dodecyl sulfate (SDS) in aqueous solution were studied using fluorescence spectroscopy. Pyrene labeled PAA was used to study changes in conformation of the polymer on its association with SDS. Externally added pyrene was used to study the aggregation behavior of the surfactant in the absence and presence of PAA. Effects of polymer concentration and pH were determined. It was observed that PAA and SDS interact, under acidic pH conditions, when the PAA concentration is low. Under conditions where the PAA was ionized, there were no interactions between PAA and SDS, but if the polymer concentration is high then an increase in ionic strength due to the dissociation of a large number of carboxylic acid groups resulted in aggregation of the surfactant, even at concentrations much lower than its critical micelle concentration in the absence of any polymer [73].

The solubility of electrolytes in moderately hydrophilic polymers is controlled by the macromolecular hydrophilic/hydrophobic balance [6]. The typical examples of these polymers are polyamides. For instance Nylon-6 is the combination of several isomorphous states: α- and γ-form are the most well-known. The α-form is the monolithic elementary crystalline lattice with plate stretched conformation of the chain. Crystallites, lamellae, or ordered microfibril domains are of layered structure. The γ-form is less stale and its conformation is possible after treatment of the polymer by water. The assumptions about its structure and some alternative models were reported in [74]. Under unfavorable conditions the imperfect γ^*-form also appears resembling the liquid-crystal (mesophase) state. All imperfect forms transform into perfect ones after interaction of mineral acids into the polymer. The evidence of this was observed in [75] during the treatment of polycaproamide by inorganic acids and was named γ–α transition.

However, the kinetic scheme proposed in [75] was formal and did not take into account the heterogeneous character of polymer matrix. Our viewpoint [74, 6] is that the elementary act of acid interaction with a polymer is a rupture of hydrogen bonds between amide or end groups coupled with the transfer of acid molecules in mesamorphous domain. The same act at the interface of mesamorphous and crystalline phases initiates the structural γ–α transition.

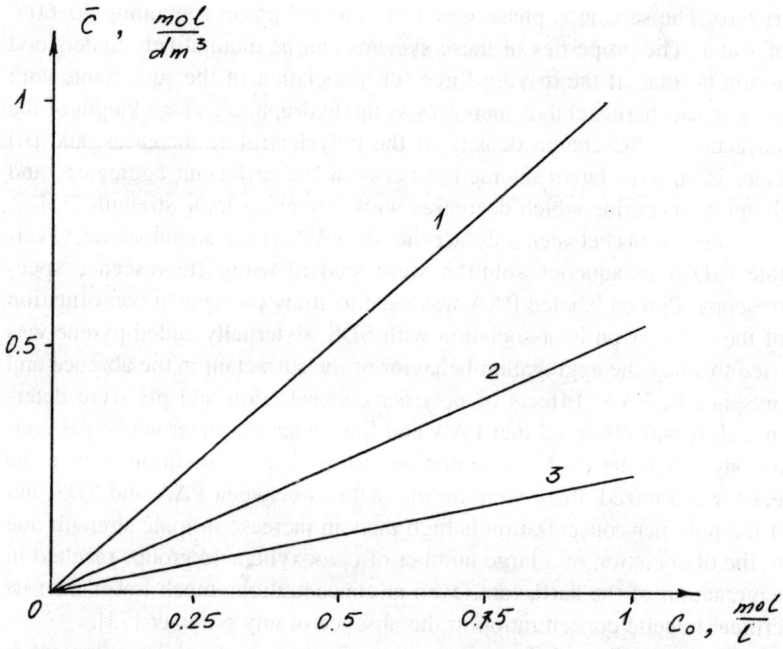

Figure 1.13. Isotherms of sorption of inorganic acids in Nylon-6: HNO3 (1), HCl (2), H2SO4 (3).

The sorption isotherms of several acids in Nylon-6 are shown in Fig. 1.13. All isotherms are linear and this proves the stability of the corresponding distribution constant K. Since the acid is distributed between amorphous and mesamorphous phases, we get

$$C_s = C_a + C_m = Kc_0 \qquad (1.17)$$

where C_s, and c_0 are acid concentration in polymer and aqueous external solution, respectively, and C_a and C_m are its concentration in amorphous and mesamorphous domains per volume unit of the polymer. The equilibrium between amorphous and mesamorphous domains is described by the reaction

$$HA_a + \nu_m \overset{K_m}{=\!=\!=} HA_m \qquad (B)$$

where HA is an acid molecule situated in amorphous (a) and mesamorphous (m) phases, ν_m is the sorption center which characterizes the mesamorphous domain, and K_m is the equilibrium constant. Following Eq. 1.17 and reaction B the distribution of acid between domains was found in terms of concentrations as

$$C_a = 0.5 [Kc_0 - A_0 - 1/K_m - (KC_0 - A_0 \pm K^2_m + 4Kc_0/K_m)^{1/2}]$$

$$C_m = 0.5 [Kc_0 + A_0 + 1/K_m - (KC_0 - A_0 \pm K^2_m + 4Kc_0/K_m)^{1/2}],$$

where A_0 is the initial concentration in immobilization centers.

Another model was proposed for analyzing sorption and volumetric behavior of glassy polymer-penetrant systems. It also takes into account the effects of structural arrangements in the polymer matrix on sorption and volumetric behavior. It was shown that the proposed theory can predict general trends for sorption and volumetric behavior in the glassy state, and furthermore, it gives good agreement with actual sorption data [76].

A detailed investigation of the sorption of 4-aminoazobenzene by cellulose acetate films from aqueous solution was reported in [77]. Sorption isotherms at 75 and 60°C were linear up to saturation, in agreement with previous findings that led to the conclusion that cellulose acetate disperse dye systems were thermodynamically ideal. Thermodynamic analysis of these data also gave results consistent with such previous findings. The isotherms for 45 and 25°C, on the other hand, exhibited increasing curvature, in line with similar recent findings for a variety of hydrophobic polymer-disperse dye systems, and are consistent with the presence of some strong absorption sites.

In hydrophobic polymers where electrolyte is usually not dissociated the distribution coefficient (Eq. 1.16) is mainly defined by the chemical contribution (K_i). This was shown, for example, of HCl vapor sorption in LDPE [78]. The formation of hydrate structure in polyethylene was reported for the sorption of aqueous HCl solution in the absence of osmotic and hydrostatic acids.

The distribution coefficients of the solutes (toluene, naphthalene, and phenanthrene) were reported at infinite dilution between silicone rubber and supercritical-fluid carbon dioxide [79]. A new technique was described in which a thin film of polymer was coated and cross-linked onto silica, and the distribution coefficient is measured rapidly by elution supercritical-fluid chromatography. Because CO_2 significantly enhances the solute's volatility and its diffusion coefficient in the polymer, it was possible to study solute-polymer interactions at room temperature for nonvolatile compounds which would be difficult to study by conventional techniques such as gas chromatography. These infinite dilution data were used to determine solute-polymer interaction parameters to calculate phase diagrams over a wide concentration range. The residual, combinatorial, and cross-link contributions to the solute activity coefficient in the polymer were discussed as a function of concentration. In addition, pronounced pressure and temperature effects were described in terms of experimentally measured solute partial molar volumes (to -14 L/mol) and partial molar enthalpies (to -850 kJ/mol) in the fluid phase.

The solubility of micromolecular solutes in rubbery polymeric media can be treated with reasonable simplicity and generality, when only nonpolar or very weakly polar interactions are involved. A regular solution approach yields a useful description of the relative solubility of different gaseous solutes, but more sophisticated treatments are necessary for the evaluation of absolute solubility. The treatment of excess solubility in glassy polymers seems to be the topic which currently attracts most interest. The dual mode sorption model, which was long used for this purpose, as well as alternative approaches recently proposed, were criticized with a number of experimental justifications in [80].

The functional dependencies of equilibrium and nonequilibrium sorption parameters on solute molecular descriptors were analyzed for 29 organic compounds and two soils. Similar correlation patterns were obtained with all three evaluated size/shape descriptors (molecular surface area, van der Waals volume, molecular connectivity). The functional dependencies of equilibrium distribution coefficients on the solute molecular descriptors were analyzed for three systems used as phenomenological models of sorption by soil (octanol, reversed-phase HPLC packing material [RPLC], polymer). The correlation patterns exhibited by the three models were compared to those reported for the soil systems. The correlation patterns exhibited by the soil data were similar to the patterns exhibited by the polymer systems and dissimilar to those exhibited by the octanol and RPLC systems. In addition, the correlation pattern between the sorption rate coefficient and molecular connectivity was similar to that between polymer-diffusion coefficients and molecular connectivity. Hence, it appears that the polymer analog may be the most appropriate of the three models for representing both equilibrium and nonequilibrium sorption by soil. Based on these results, the polymer analog was suggested as the phenomenological model of choice for investigating and evaluating the sorption dynamics of low-polarity organic compounds in soil systems [81].

The thermodynamics of solute sorption in polymeric materials defines further process of the diffusion. The effect of electrolyte on polymer and back may occur at different levels, which are (i) the molecular level; (ii) the structural and morphological level; (iii) the relaxation level; and (iv) the macroscopic level (hydrodynamic conditions) [6].

1.4. DIFFUSION OF SOLUTES IN POLYMERS

The diffusion of solutes in polymers is almost impossible to separate from the diffusion of solvent, mainly of water. The simultaneous transport of water and solute is the topic of chapter 2. Here we would like just to remind the reader of the main theoretical approaches to the description of transport of low-molecular solutes in different kinds of polymers. Obviously, we selected models which are more or less justified experimentally.

It is convenient to begin the consideration of the diffusion of solute in polymers with the theory of steric hindrance of McKay-Meares [82] and the free volume theory of Yasuda [83]. Both of them relate the diffusion coefficient of the solute in polymer (D_s) to its diffusion coefficient in aqueous solution (D_{s0}).

The McKay-Meares theory explains the reduction in self-diffusion coefficients of low-molecular compounds in polymers by elongation of the diffusion path in comparison with aqueous solution. If ϕ_p is the volume fraction of polymer, this assumption leads to the expression

$$D_s = D_{s0}[(1-\phi_p)/(1+\phi_{sp})].$$

The equation remains valid only for moderate concentrations of solute and low ϕ_p. Attempts were made to extend the application of this model by introduction of a correlation parameter which was the ratio between the viscosity in polymer (ν_p) and in aqueous solution (ν) [84].

$$D_s = D_{s0}(\nu_p/\nu)[(1-\phi_p)/(1+\phi_{sp})]$$

However, no physical explanation for this correction had been done.

In contrast to the previous theory, the "free volume theory" of Yasuda and Peterlin is based on the following assumptions:

(i) the effective free volume which accommodates the dissolved compound is equal to the free volume of water in the polymer;

(ii) the size of voids of fluctuation which make up the free volume must be co-sized with the diffusing particle; and

(iii) there is no interaction between the molecules of polymer and solute.

For these assumptions the diffusion coefficient of solute in the water-containing polymer may be given as

$$D_s = D_{s0}\exp[(-BS_s(\phi_w^{-1}-1)/f^w_v] \tag{1.18}$$

where S_s is the cross-sectional area of the diffusing particle, ϕ_w and f^w_v

are volume fraction of water in the polymer and its free volume, respectively. Therefore, the acceleration of the solute diffusion in more hydrophilic polymers as well as the reduction of the diffusion coefficient with the increase of the radius of solute molecule should be expected. Experimental evidence of this fact is discussed in detail in [6]. The equation may be rewritten in terms of the concentration of water in polymer (C_w)

$$D_s = D_{s0}exp[(-k_{w3}/(C_w \mp k_{w4})]$$ (1.19)

where the meaning of constants can be understood by the comparison of Eqs. 1.18 and 1.19.

Free volume theory does not take into account dissociation and, therefore, should be applied for neutral solutes. It also does not describe the diffusion in polymers which contain charged groups [6]. The modern development of free volume theory was reported by Duda and Vrentas who gave comprehensive analysis of transport of low-molecular weight compounds above [85] and below [86] glass transition temperature.

The diffusion of electrolytes in polymers containing ionogenic groups was considered by Kokotov [87], Nilolayev [88], Ijima [89, 90], Timashev [91], and other researchers. This problem is discussed in greater detail in [6].

Recently, the dependencies of the electroosmotic permeability on concentration and the water content of ion-exchange membranes were investigated together with the dependencies of the conductivity and the ion transport numbers over a wide range of sodium chloride concentrations. The materials tested were samples of heterogeneous membranes based on ion-exchange resins and samples of homogeneous perfluorinated membranes. Theoretical description of the results was performed with the help of a model approach regarding a membrane as a system consisting of two conventional conductive phases. The final model equation for the water transport number showed their correlation with the physicochemical characteristics of the polymer material and enables the evaluation of the hydration parameters of the ion-dipole complexes in the cluster phase of the membrane [92].

The main problem which occurs when we study the solute diffusion in polymer is that for different reasons electrolyte does not diffuse uniformly. The mechanism of the diffusion depends, for example, on whether the electrolyte in the polymer exists in dissociated form or as neutral molecule. In the case of acid the dissociation leads to the appearance of the fastest form, protons, in the system. For dissociating electrolytes the dependence of the diffusion coefficient on ion radius is affected by the dis-

Table 1.2. Diffusion coefficients for chlorine-substituted acetic acid in PVA at 25°C and $\phi_w = 0.67$

Acid	Anion R, A	$K_D \times 1000$	$D_s \times 10^6$ cm^2/s
CH$_3$COOH	2.25	0.0175	1.8
CH$_2$ClCOOH	2.41	1.38	2.6
CHCl$_2$COOH	2.55	19.0	3.0
CCl$_3$COOH	2.67	200	3.5

sociation degree. Madyuskin found [93] that for a number of chlorine substituted aliphatic acids, the effect of the constant of dissociation (K_D) on diffusion coefficient in PVA is more essential than the effect of anion radius (Table 1.2). In the meantime, he reported the expected dependence of D_s on radius for formic, acetic, propionic, and butyric acids [93].

For less (so-called, moderately) hydrophilic polymers another problem becomes the most important. This is the consequence of the heterogeneous character of polymer structure. For these polymers the total concentration of solute, C_s, at any point in the matrix was composed of a mobile or "dissolved" part and an immobile or "dispersed" part [94]. The lamellas, crystallites, spherulites, and other ordered structures are not permeable for diffusing components of solutes. Peterlin proposed to characterize the polymer crystalline phase by two parameters: crystallinity degree (α_c) and orientation of crystalline structure in polymer [95]. The diffusion coefficient is reduced and assumed equal to

$$D_s = (1-\alpha_c)_{sas}^m$$

where D_{as} is the diffusion coefficient in amorphous phase and m is the empirical parameter (m $> 0.3/1.9$).

The kinetic analysis of the sorption of 4-aminoazobenzene by cellulose acetate films [77] has shown that the situation appears to be more complex than envisaged by simple dual-mode sorption theory. Early time desorption kinetic data were found to be consistent with Fick's law, with no indication of any significant dependence of the diffusion coefficient D on concentration (in the medium- to high-concentration range) or on film thickness. On the other hand, D was affected significantly by the history of film formation, the method of introducing the dye or heat treatment of the film. The origin of these effects was traced by DSC to definite microstructural differences. Late time kinetic data deviated significantly from the theoretical predictions based on the corresponding early time data, indicating a progressive slowdown of the desorption process. The deviations in question were generally more marked at higher temperature or film thickness, or when dyeing had been effected from the vapor phase, and were attributed to slow release of strongly adsorbed dye molecules.

Nylon-6 is a typical example of moderately hydrophilic polymers with structural heterogeneity. In the previous paragraph we considered some peculiarities of the acid sorption in this polymer. This would be also a good example to show the effect of polymer structure on the diffusion of solutes.

The analysis of the results of acid interaction with polyamides showed [74] that the model of diffusion should take into account the following regularities:

(i) the equilibrium at the polymer–surface interface is reached according to the mechanism which is described by Eq 1.17;

(ii) the maximal concentration of the acid in mesamorphous domain does not depend on its concentration in aqueous solution; and

(iii) the diffusion of acid in mesamorphous phase is accompanied by an irreversible reaction of the rupture of hydrogen bonds between amide groups.

With these statements, acid diffusion in the amorphous domain was described by the equation

$$\delta C_{as}/\delta t = D_{as}\delta^2 C_{as}/\delta x^2.$$

In the mesamorphous domain, the transport is accompanied by the reaction of hydrogen rupture and is described by equation

$$\delta C_{ms}/\delta t = D_{ms}\delta^2 C_{ms}/\delta x^2 - kC_{as}(A_0 - C_{as})$$

where A_0 is the initial concentration of amide groups sutured by hydrogen bonds, D_{as} and D_{ms} are the diffusion coefficients in the mesamorphous phase, respectively, and k is the constant of the reaction of rupture of hydrogen bonds.

The application of the above model gave us the opportunity to explain the experimental results. Particularly for the boundary conditions

$$C_{as}(0,t) = C_{a0}; \quad C_{ms}(0,t) = C_{m0};$$

$$C_{as}(x,0) = C_{ms}(x,0) = C_{as}(l,t) = C_{ms}(l,t) = 0; \quad C_{a0} + C_{m0} = Kc_0;$$

we found the expressions for a steady-state flux (J) and the permeability (P) of acid in Nylon-6.

Table 1.3. Parameters of sorption and diffusion of inorganic acids in Nylon-6

Parameters	Nitric	Nitric*	Acid Hydrochloric	Sulfuric
$D_{as} \times 10^8$ cm^2/s	2.3	2.3	1.6	0.8
$D_{ms} \times 10^8$ cm^2/s	1.7	1.7	0.6	0.6
$k \times 10^{-2}$ s^{-1}	0.36	0	0.63	0.17
K dm^3/l	1.08	1.08	0.58	0.22
A_0 mol/l	0.32	0.32	0.32	0.16
K_m l/mol	3.0	3.0	3.0	3.0
D_{ms}/D_{as}	0.7	0.7	0.4	0.75

* Polymer was preliminary treated with HNO$_3$ for 4 hours.

$$J = D_{as}C_{a0}/l + (D_{as} kA_0)^{1/2}C_{m0}/\sinh(K/D_{ms})$$
$$P = D_{as}kA_0c_0 - D_{as}(1 - D_{ms}y/(D_{as}\sinh y))C_{m0}$$

where $y = (kl^2/D_{ms})^{1/2}$; $C_{m0} = K_mA_0C_{a0}/K_mC_{a0} + 1$.

The parameters of the transport of inorganic acids in Nylon-6 are collected in Table 1.3.

The proposed diffusion model helped also to explain the kinetics of structural rearrangement in Nylon-6 [74].

Sorption and transport of aqueous salt solutions in polyurethane elastomer was investigated over the temperature interval of 25–60°C by using an immersion/weight gain method. The size of the ions has no significant affect on the penetrant transport coefficients. The sorption data have been explained in terms of the simple Fickian model. Attempts also were made to estimate the transport parameters by applying corrections to include the spatial diffusivity of the liquids within the polymer sample. Temperature dependence of the transport coefficients was used to estimate the activation parameters from the Arrhenius plots [96].

Although an application of pure hydrophobic polymers (polyolefins, fluoroplastics, and others which take no more than 0.1% of water in equilibrium) in systems of controlled release is very limited, the regularities of electrolyte transport in these polymers are important for further understanding of the behavior of solutes in moderately hydrophobic materials, especially in polymer blends and composites.

The substantial difference in flows of different electrolytes through the same polymer is one of the most general features in electrolyte transport in hydrophobic polymers [78]. Such selectivity is not observed in more hydrophilic polymers although there may be some difference in the rate of diffusion. In hydrophobic polymers this difference is of several orders of magnitude. For example, the penetration of hydrogen chloride across a LDPE film can be detected minutes after the beginning of the process whereas KCl penetration could not be detected even 3 months later. Penetration of nitric acid across PET films was observed after a few tenths of a minute, and more than 1 year of exposure was required for sulfuric acid [97]. The effect of the nature of electrolyte on the rate of its diffusion may be, probably, explained by the dependence of the diffusion coefficient of electrolyte on the size of its molecule [78]. Unfortunately, this assumption still gets insufficient experimental support.

The dissolution rates of sparingly soluble, fine particulate, suspended drugs have been studied using a Coulter Counter Model TAII. For two sieve fractions of oxazepam the dissolution rates were monitored in media with varying viscosity brought about by the addition of glycerol, while for griseofulvin the change in the medium's viscosity was induced by changing the temperature. By calculating the dissolution rate, and compensating for differences in particle surface area and media solubility, it was shown that the dissolution rate was diffusion controlled. After additional normalization for the diffusion coefficient, it was suggested that the so-called apparent diffusional distance decreased substantially with particle size. The effect of particle size was more limited above approximately 15 μ [98].

The second contributing factor in the diffusion coefficient (D_s) of electrolyte in hydrophobic polymer is the temperature. Many authors [78, 99–101] noted the exponential D_s dependence on the reciprocal of absolute temperature.

The acceleration of the diffusion of electrolytes with the increase of its concentration in external aqueous solution was shown for many systems. The so-called acid penetration factor λ_D ($D_s = K_1\lambda^2_D$) was used [78] to compare the rate of diffusion. The empirical formula for the concentration dependence of λ_D was proposed in [102]

$$\lambda_D = ap^b$$

where p is the vapor pressure of electrolyte above the solution (activity of solute in general case), and a and b are empirical constants. For a number of systems b was found equal to 0.5. In other words, the linear dependence of the diffusion coefficient of solute on its activity in external phase was assumed. The concentration dependence of D_{HCl} was studied in [103] and explained through the existence of two forms of solute in polymer. These are undissociated molecules which are involved in mass transfer and complexes with water of very low mobility. This model was developed by Markin [78] who postulated that the only water-solute interaction can provoke non-Fickian solute diffusion. This is of course if chemical reactions are not considered. The following equations were proposed to relate the parameters of solute diffusion and the composition of the external solution

$$[\delta(\ln D_s)/\delta n)]_T = -\Delta m_w /RT$$

$$[\delta(\ln \lambda_D)/\delta n)]_T = d_0 \ln a_w$$

$$\lambda^2_D = r_0 [a_s (a_w)^n]^{d_0}$$

where D_s is the diffusion coefficient of solute in polymer; n is the number of moles water per mole of solute in external solution; μ_w is the difference in chemical potentials of water in aqueous solution and in pure water; a_w and a_s are the activities of water and solute, respectively; r_0 and d_0 are the phenomenological constants.

Although the application of this model is rather limited by certain polymer-electrolyte systems we should emphasize the finding that the diffusion in hydrophobic polymers is better described in terms of activities than concentrations.

Parker and Ranney examined sorption of low ppb levels of organic solutions by polytetrafluoroethylene (PTFE), rigid polyvinyl chloride (PVC), and stainless steel 304 and 316 well casings. Nineteen organics were selected, including several munitions and chlorinated solvents. Compounds were selected to offer a range of physical properties, such as solubility in water, octanol/water partition coefficient, and molecular structure. When these results were compared with the results from a similar study conducted at ppm levels, the rate and extent of sorption by PTFE and PVC were the same as seen previously for almost all analytes. There were no losses of any compounds associated with stainless steel. At these low levels (ppm and ppb), the rate of diffusion within the polymer (PVC and PTFE) is independent of concentration. Only with PTFE are the rates rapid enough to be of concern when monitoring for some contaminants in ground water. Tetrachloroethylene was the compound PTFE sorbed the most rapidly. The study showed that PVC well casings are suitable for monitoring low levels (ppm and ppb) of organics [104].

The migration of several low molecular weight model solutes through a polypropylene/polyolefin blend was examined via a permeation-cell experimental design. Diffusion coefficients and permeation rates are obtained for the model solutes. The permeation rates exhibit Arrhenius-type behavior. A bimodal permeation model was proposed wherein the permeation rate is correlated with the linear combination of a solute's Permicor constant and solute/polymer equilibrium interaction constant. The model effectively predicts the behavior of the polymer and solutes studied. The implication of the observed behavior to container compatibility assessment was discussed [105].

In [78] it was proposed that apart from normal diffusion mechanism some kind of "slow" diffusion occurs in the hydrophobic polymers. Normally, its contribution in the overall transport is negligible. However, at low solubility of electrolyte in the polymer or at its low activity the "slow" diffusion may be observed. The mechanism of this "slow" diffusion received so far little discussion. It is assumed that the rate of this process weakly depends on the nature of solute and its concentration in aqueous solution.

We can see from the above that even in hydrophobic polymers the effect of water on the diffusion of solute is essential. The aim of the following chapters is to consider the simultaneous transport of water and solutes in detail.

REFERENCES

[1] B.A. Firestone, R.A. Siegel. *J. Applied Polym. Sci.,* **43**, 901 (1991).

[2] W.P. Hsu, A.S. Myerson, T.K. Kwei. *Europ. Polym. J.,* **29**, 1601 (1993).

[3] H.J.V. Tyrell. *J. Chem. Educ.,* **41**, 397 (1964).

[4] V.N. Manin, A.N. Gromov. *Physicochemical Stability of Polymeric Materials in Conditions of Exploitation.* Khymiya, Moscow (1980). (In Russian).

[5] D.W. McCall et al. *Macromolecules,* **17**, 1644 (1984).

[6] A.L. Iordanskii, T.E. Rudakova, G.E. Zaikov. *Interaction of Polymers with Bioactive and Corrosive Media.* VSP, Utrecht, the Netherlands (1994).

[7] J.L. Williams, H.B. Hopfenberg, V. Stannet. *J. Macromol. Sci. Phys.,* **3**, 711 (1969).

[8] S. Brunauer, P.H. Emmett, E. Teller. *J. Am. Chem. Soc.,* **60**, 309 (1938).

[9] P.J. Flory. *Principles of Polymer Chemistry.* Cornell University Press, New York (1953).

[10] J.L. Lundberg. *J. Macromol. Sci. Phys.,* **3**, 611 (1969).

[11] B.H. Zimm, J.L. Lundberg. *J. Phys. Chem.,* **60**, 425 (1956).

[12] J.A. Barrie, B. Platt. *Polymers,* **4**, 303 (1963).

[13] Yu.S. Zuev. *Degradation of Polymers under the Action of Aggressive Media.* Khimiya, [Moscow (1972).

[14] J.I. Rogers (Ed.). *Permeability of Plastic Films and Coats.* Academic Press, New York (1974).

[15] G.L. Brown. In: *Water in Polymers.* S.P. Roland (Ed.), p. 451, ACS Symposium Series 127 (1980).

[16] A.Ya. Polishchuk, A.L. Iordanskii, G.E. Zaikov, A.M. Ivanov, V.A. Murov. *Plasticheskiye massy (Plastics),* No. 11, 38 (1988) (In Russian).

[17] T.A. Orofino, H.B. Hopfenberg, V.Stannet. *J. Macromol. Sci. Phys.,* **3**, 778 (1969).

[18] J.A. Barrie, D. Machin. *J. Macromol. Sci. Phys.,* **3**, 645 (1969).

[19] R.J. Hernandez, R. Gavara. *J. Polym. Sci. Polym. Phys.,* **32**, 2367 (1994).

[20] M.C. Lee, N.A. Peppas. *J. Appl. Polym. Sci.,* **47**, 1349 (1993).

[21] Y. Tamai, H. Tanaka, K. Nakanishi. *Macromolecules,* **27**, 4498 (1994).

[22] F. Kawano, E.R. Dootz, A. Koran, R.G. Craig. *J. Prosthetic Dentistry,* **72**, 393 (1994).

[23] J. Rapply, J.Yang and G. Tollin. In: *Novel Instrumental Methods for Polymer Structure Investigations,* p. 114, Mir, Moscow (1982).

[24] I.C. McNeill, A.Ya. Polishchuk, G.E. Zaikov. *Polym. Degrad. Stab.,* **37**, 223 (1992).

[25] J.T. Hinatsu, M. Mizuhata, H. Takenaka. *J. Electrochem. Soc.,* **141**, 1493 (1994).

[26] P.I. Lee. *J. Appl. Polym. Sci.,* **42**, 3077 (1991).
[27] J.H. Petropoulos. *J. Membr. Sci.,* **18**, 37 (1984).
[28] J. Komiyama et al. *Polym. J.,* **23**, 379 (1991).
[29] R.J. Hernandez. *J. Food Eng.,* **22**, 495 (1994).
[30] M.E. McNeill, N.B. Graham. *J. Biomed. Sci. Polym. Ed.,* **4**, 305 (1993).
[31] G. Khamrakulov, N.V. Myagkova, V.P. Budtov. *Vysokomolekulyarnye soedineniya,* **36B**, 845 (1994).
[32] N.E. Schlotter, P.Y. Furlan. *Polymer,* **33**, 3323 (1992).
[33] A.E. Chalyh. *Diffusion in Polymeric Systems,* Khimiya, Moscow (1987) (In Russian).
[34] W.P. Hsu, R.J. Li, A.S. Myerson, T.K. Kwei. *Polymer,* **34**, 597 (1993).
[35] S.B. Lee, T.J. Rockett, R.D. Hoffman. *Polymer,* **33**, 3691 (1992).
[36] B. Deneve, M.E.R. Shanahan. *Polymer,* **34**, 5099 (1993).
[37] M.P. Patel, M. Braden. *Biomaterials,* **12**, 653 (1991).
[38] M.E. Best, C.R. Moylan. *J. Appl. Polym. Sci.,* **45**, 17 (1992).
[39] N.A. Peppas, L. Brannon-Peppas. *J. Food Eng.,* **22**, 189 (1994).
[40] J.S. Smith, N.A. Peppas. *J. Appl. Polym. Sci.,* **43**, 1219 (1991).
[41] S.H. Gehrke, D. Biren, J.J. Hopkins. *J. Biomaterials Sci. Polym. Ed.,* **6**, 375 (1994).
[42] J.C. Wu, N.A. Peppas. *J. Polym. Sci. Polym. Phys.,* **31**, 1503 (1993).
[43] J.C. Wu, N.A. Peppas. *J. Appl. Polym. Sci.,* **49**, 1845 (1993).
[44] J.H. Petropoulos, P.R. Roussis. *J. Membr. Sci.,* **3**, 343 (1978).
[45] J.H. Petropoulos. *J. Polym. Sci. Polym. Phys. Ed.,* **22**, 183 (1984).
[46] A.Ya. Polishchuk, G.E. Zaikov, J.H. Petropoulos. *Inter. J. Polymer. Mater.,* **25**, 1 (1994).
[47] A.Ya. Polishchuk, G.E. Zaikov, J.H. Petropoulos. *Inter. J. Polymer. Mater.,* **19**, 1 (1993).
[48] A.L. Iordanskii, A.Ya. Polishchuk, L.P. Razumovskii. *Indian J. Chem.,* **31A**, 366 (1992).
[49] A.Ya. Polishchuk, L.A. Zimina, R.Yu. Kosenko, A.L. Iordanskii, G.E. Zaikov. *Polym. Degrad. Stab.,* **31**, 247 (1991).
[50] M.A. Delnobile, G. Mensitieri, P.A. Netti, L. Nicolais. *Chem. Eng. Sci.,* **49**, 633 (1994).
[51] J.L. Duda, I.H. Romdhane, R.P. Danner. *J. Non-Crystalline Solids,* **172**, 715 (1994).
[52] J.M. Zielinski, J.L. Duda. *AICHE J.,* **38**, 405 (1992).
[53] P.I. Lee, C.J. Kim. *J. Membr. Sci.,* **65**, 77 (1992).
[54] N.J.M. Kuippers, A.A.C.M. Beenackers. *Chem. Eng. Sci.,* **48**, 2957 (1993).
[55] M. Best, J.W. Halley, B. Johnsson, J.L. Valles. *J. Appl. Polym. Sci.,* **48**, 319 (1993).
[56] F. Bellucci, L. Nicodemo. *Corrosion,* **49**, 235 (1993).
[57] J.H. Jou, R. Huang, P.T. Huang, W.P. Shen. *J. Appl. Polym. Sci.,* **43**, 857 (1991).
[58] C.A. Fyee, L.H. Randall, N.E. Burlinson. *J. Polym. Sci. Polym. Chem.,* **31**, 159 (1993).
[59] D.R. Morris, X.D. Sun. *J. Appl. Polym. Sci.,* **50**, 1445 (1993).
[60] F. Debeaufort, A. Voilley, P. Mearce. *J. Membr. Sci.,* **91**, 125 (1994).
[61] V.S. Markin, G.E. Zaikov. *Polym. Degrad. Stab.,* **40**, 395 (1993).
[62] M. Pisarik, L. Lapcik. *Chemcial Papers - Chemicke Zvesti,* **47**, 19 (1993).
[63] M. Sanopoulou, J.H. Petropoulos. *J. Polym. Sci. Polym. Phys.,* **30**, 983 (1992).
[64] N.A. Peppas, J.C. Wu, E.D. Vonmeerwall. *Macromolecules,* **27**, 5626 (1994).
[65] M.C. Lee, N.A. Peppas. *J. Composite Mater.,* **27**, 1146 (1993).
[66] J.A. Horas, F. Nieto. *J. Polym. Sci. Polym. Phys.,* **32**, 1889 (1994).
[67] K.I. Okamoto et al. *J. Polym. Sci. Polym. Phys.,* **30**, 1223 (1992).
[68] I. Saucedo, E. Guibal, C. Roulph, P. Lecloirec. *Environmental Technology,* **13**, 1101 (1992).
[69] N. Lakshminarayanaiah. *Transport Phenomena in Membranes.* Academic Press, New York (1969).
[70] Yu.A. Kokotov, P.P. Zolotarev, G.E. El'kin. *Theoretical Ground of Ion Exchange.* Khimiya, Leningrad (1986) (In Russian).

[71] L.A. Zimina, N.N. Madyuskin, A.Ya. Polishchuk, A.L. Iordansky, G.E. Zaikov. *Inter. J. Polymer. Mater.*, **16**, 185 (1992).
[72] M. Barck, P. Stenius. *Colloids and Surfaces. Physicochem. Eng. Aspects*, **89**, 59 (1994)
[73] C. Maltesh, P. Somasundaran. *Colloids and Surfaces*, **69**, 167 (1992).
[74] A.L. Iordansky, A.Ya. Polishchuk, R.Yu. Kosenko et al. *Inter. J. Polymer. Mater.*, **16**, 195 (1992).
[75] A.N. Machyulis, A.W. Kwiklis, E.E. Tornau et al. *Trudy Litovskoi Akademii Nauk (Proceedings of the Lithuanian Academy of Sciences)*, **1**, 151 (1971).
[76] J.S. Vrentas, C.M. Vrentas. *Macromolecules*, **24**, 2404 (1991).
[77] K.G. Papadokostaki, J.H. Petropoulos. *J. Polym. Sci. Polym. Phys.*, **32**, 2347 (1994).
[78] G.E. Zaikov, A.L. Iordanskii, V.S. Markin. *Diffusion of Electrolytes in Polymers*, VSP, Utrecht, The Netherlands (1988).
[79] J.J. Shim, K.P. Johnston. *AICHE J.*, **37**, 607 (1991).
[80] J.H. Petropoulos. *Pure and Appl. Chem.*, **65**, 219 (1993).
[81] M.L. Brusseau. *Environmental Toxicology and Chemistry*, **12**, 1835 (1993).
[82] D. McKay, P. Meares. *Trans. Faraday Soc.*, **55**, 1221 (1959).
[83] H. Yasuda, C.E. Lamaze, D. Ikenberry. *Macromol. Chem.*, **118**, 19 (1968).
[84] A. Despic, J. Hills. *Trans. Faraday Soc.*, **53**, 1262 (1957).
[85] J.S. Vrentas, J.L. Duda, A.C. Hou. *J. Polym. Sci. Polym. Phys. Ed.*, **23**, 2469 (1985).
[86] J.S. Vrentas, J.L. Duda, H.C. Ling. *J. Polym. Sci. Polym. Phys. Ed.*, **26**, 1059 (1988).
[87] Yu.A. Kokotov, V.A. Pasechnik. *Equilibrium and Kinetics of Ion Exchange*, Khimiya, Leningrad (1970) (In Russian).
[88] N.I. Nikolaev. *Diffusion in Membranes*, Khimiya, Moscow (1980) (In Russian).
[89] T. Ijima, T. Obara, M.Isshiku et al. *J. Colloid. Interface Sci.*, **63**, 1211 (1978).
[90] Y. Kawana, J. Komiyama, J.H. Petropoulos, T. Ijima. *J. Polym. Sci. Polym. Phys. Ed.*, **23**, 1813 (1985).
[91] S.F. Timashev, A.A. Ovchinnikov, A.A. Belyi. *Kinetic Diffusion-Controlled Chemical Processes*. Khimiya, Moscow (1987) (In Russian).
[92] N. Berezina, N. Gnusin, O. Dyomina, S. Timofeev. *J. Membr. Sci.*, **86**, 207 (1994).
[93] N.N. Madyuskin. *Thesis for the Degree of Candidate of Science (PhD Thesis)*, Institute of Chemical Physics, the USSR Academy of Sciences, Moscow (1991) (In Russian).
[94] T. Higuchi. *J. Pharm. Sci.*, **52**, 1145 (1963).
[95] A. Peterlin. *J. Separ. Sci. Technol.*, **14**, 139 (1979).
[96] S.B. Harogoppad, T.M. Aminabhavi, R.H. Balundgi. *J. Appl. Polym. Sci.*, **42**, 1297 (1991).
[97] D.D. Chegodaev, Z.K. Naumova, Ts.S. Dunaevskaya. *Fluoroplastics*, GKhI, Leningrad (1960) (In Russian).
[98] M. Bisrat, E.K. Anderberg, M.I. Barnett, C. Nystrom. *Inter. J. Pharmaceutics*, **80**, 191 (1992).
[99] V.I. Ryasantsev, M.F. Sisin, K.S. Minsker. *Plasticheskiye Massy (Plastics)*, No. 4, 33 (1975) (In Russian).
[100] N.S. Tikhomirova, K.I. Zernova, V.N. Kotrelev. *Plasticheskiye Massy (Plastics)*, No. 4, 33 (1963) (In Russian).
[101] M. Mozisek. *Plasty Kauc.*, **15**, 1 (1978).
[102] V.A. Murov. *Thesis for the Degree of Candidate of Science (PhD Thesis)*, Moscow Institute of Chemical Industry (1970) (In Russian).
[103] A.L. Shterenzon, S.A. Reitlinger, L.P. Topina. *Polymer Sci. USSR*, **9A**, 887 (1969).
[104] L.V. Parker, T.A. Ranney. *Ground Water Monitoring and Remediation*, **14**, 139 (1994).
[105] D.R. Jenke. *J. Appl. Polym. Sci.*, **44**, 1223 (1992).

2 GENERAL REGULARITIES AND MODELS OF SIMULTANEOUS TRANSPORT OF SOLVENT AND SOLUTES IN POLYMERS

2.1. HYDROPHILIC POLYMERS

The general principles of simultaneous transport of water and low-molecular weight solutes in hydrophilic polymer include the consideration of the processes of water diffusion in and solute diffusion from the polymer matrix, the increase of volume of the test piece (swelling) and change in mechanical properties of the polymer.

The simplest mathematical description of these processes was proposed by Peppas and coauthors [1]. In assumption of constant diffusion coefficients of both components (D_w, D_s), they wrote

$$\delta C_w/\delta t = D_w \delta^2 C_w/\delta x^2 \qquad 0 < x < 1 \qquad (2.1)$$

$$\delta C_s/\delta t = D_s \delta^2 C_s/\delta x^2 \qquad 0 < x < 1 \qquad (2.2)$$

where C_w, C_s are concentrations of water and solute in polymer, respectively. During the time of swelling t the thickness of sample increases from the initial value l_0 to the current value l, and the front of solvent penetrates to the depth X(t) (Fig. 2.1). This defines certain initial and boundary conditions which are

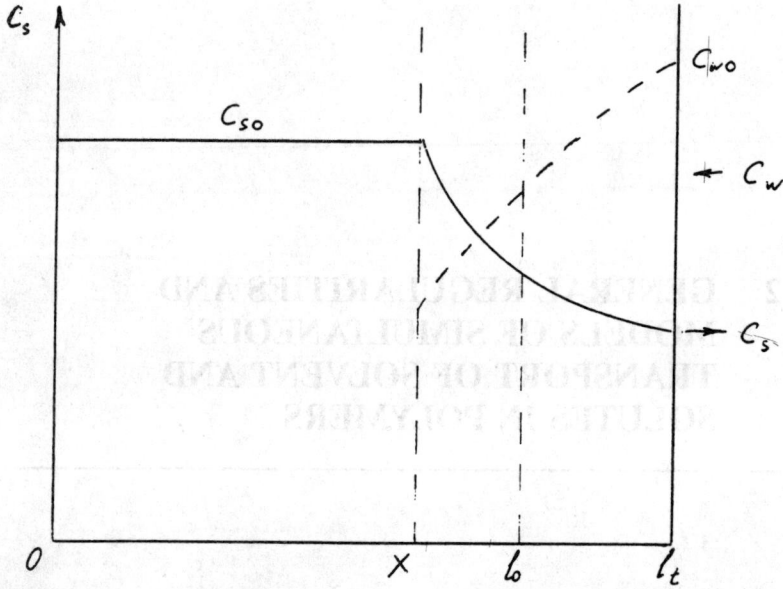

Figure 2.1. Concentration profile in a polymeric system for controlled release.

$$C_s(X,t) = C_{s1}; \quad C_s(l,t) = 0; \quad C_w(l,t) = C_{w0}$$

$$\delta C_s(0,t)/\delta x = 0; \quad \delta C_w(X,t)/\delta x = 0$$

$$C_s(x,0) = C_{s0}; \quad C_w(x,0) = 0$$

where C_{w0} is the water concentration in polymer in equilibrium, C_{s0} is the initial concentration of solute in polymer. The physical meaning of C_{s1} is less clear although its relation to the solubility of solute in polymer and water is obvious.

The above model can reproduce some of the known kinetic curves of the solute desorption. One such case occurs when the kinetics of release follows water uptake ($D_w << D_s$) [2].

The phenomenological approaches give also an opportunity to get analytical solutions of diffusion equations [2, 3] and calculate parameters which allow us to compare the kinetics and rate of the transport of components in polymer. We also applied one of these approaches in an attempt to describe the simultaneous transport of water and dioxidine in copolymers of vinylpyrrolidone with butyl (VP/BM) and methyl (VP/MM) methacrylate. Then [4] we assumed that water sorption follows the equation

$$\delta C_w/\delta t = (\delta/\delta x)(D_w \delta C_w/\delta x) - v \delta C_w/\delta x$$

where v is the velocity of local volume transfer, controlled by swelling stresses in the matrix.

Although we got reasonable agreement between theoretical and experimental results, and could make qualitative conclusions regarding regularities of the behavior of the aforesaid systems, we could not extend this model quantitatively to other systems. This is the main disadvantage of phenomenological approaches as a whole.

Therefore, the analysis of real processes in swelling matrix is required for explanation of the behavior of a system hydrophilic polymer–water–solute. The main processes that affect such behavior are considered below.

2.1.1. Effect of the mechanical properties of the polymer

The change in a stage from initial to new equilibrium may be characterized by an average relaxation time τ. Comparison of this parameter with the characteristic diffusion time (ν_D) allows us to classify polymer–solvent systems.

Vrentas et al. [5] introduced the dimensionless Deborah number (De) for comparative analysis of processes of diffusion and relaxation.

$$De = \tau/\nu_D$$

where τ depends on the elastic module and $\nu_D = R^2/D_w$, where R is the so-called characteristic length of the test piece. If De is small enough (De < 0.1), the relaxation (either mechanical or structural) is completed at an early stage of diffusion process. For this case the diffusion can be described by equations in standard form. For higher De values (De > 10) the structural rearrangement does not occur, and solvent diffusion remains Fickian. The contribution of relaxation in diffusion becomes remarkable at De values of order of 1 when the transport of solvent molecules is accompanied by changes in macromolecular conformation. Fig. 2.2 shows diagrams of Deborah numbers which were reported [2] for the ethylbenzene-polystyrene system in a finite interval of temperature and concentration. The authors emphasized the existence of elastic (De >> 1), viscous (De << 1) and viscoelastic (De ~1) diffusion. The latter is usually named anomalous, i.e., non-Fickian.

We mentioned in chapter 1 the model which was developed by Wu and Peppas [6, 7] in order to explain the anomalous penetrant diffusion behavior in glassy polymers. There the Deborah number was proposed as the

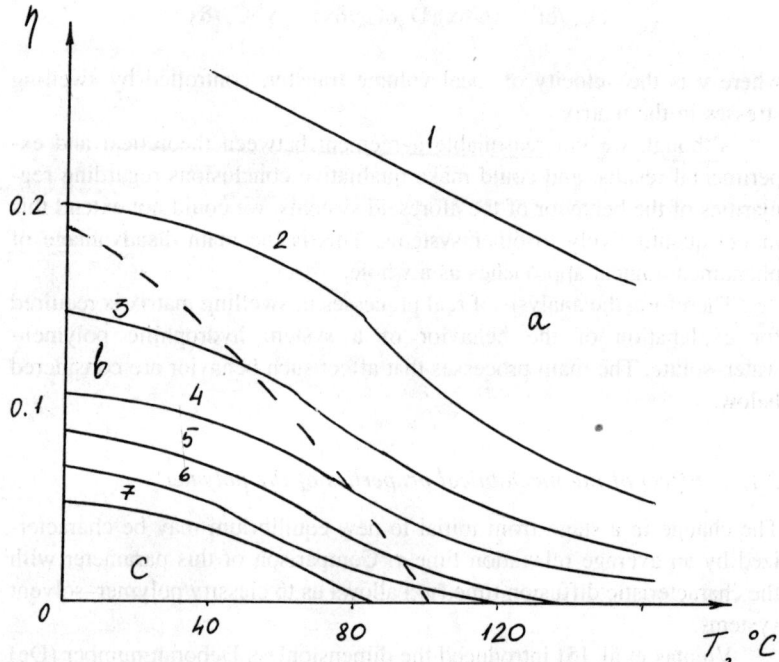

Figure 2.2. Diagram of Deborah numbers for system polystyrene-ethylbenzene. Mechanical behavior of the polymer is viscous (a), viscoelastic (b), and elastic (c). De = 0.001 (1), 0.01 (2), 0.1 (3), 1 (4), 10 (5), 100 (6), 10,000 (7). The dotted line corresponds to the T_g decrease.

ratio of the characteristic relaxation time in the glassy region to the characteristic diffusion time in the swollen region. This definition allows us to describe an integral sorption process by a single Deborah number which becomes a major parameter affecting the transition from Fickian to anomalous diffusional behavior.

The transient dynamic swelling and dissolution behavior during drug release from hydroxypropylmethyl cellulose (HPMC) matrices was investigated using fluorescein as a model drug. A new flow-through cell capable of providing a well-defined hydrodynamic condition and a nondestructive mode of operation was designed for this purpose to assess the associated moving front kinetics. The results obtained showed a continuous increase in transient gel layer thickness irrespective of the polymer viscosity grade or drug loading. This was attributed to the faster rate of swelling solvent penetration than that of polymer dissolution under the present experimental condition. On the other hand, the observed shrinkage of sample diameter over a longer time period demonstrated that polymer dissolution occurred in HPMC matrices. Further, both the rates of polymer swelling and dissolution as well as the corresponding rate of drug release increased with either higher levels of drug loading or lower viscosity grades of HPMC. For water-soluble drugs, the present results suggested that the effect of HPMC dissolution on drug release is insignificant, and the release kinetics were mostly regulated by a swelling-controlled diffusional process, particularly for higher viscosity grades of HPMC [8].

The Petropoulos model of "differential swelling stress" [9, 10], which was described in chapter 1, was generalized [11, 12] for a case of multi-component transport of solvent and solute in swelling polymers. We should repeat that this model describes the transport of solvent by the diffusion equation of a standard form

$$\delta C_w/\delta t = (\delta/\delta x)(D_w \delta C_w/\delta x) \qquad 0 < x < 2l \qquad (2.3)$$

where

$$D_w = D_w exp(k_{w1}C_w + k_{w2}f) \qquad (2.4)$$

coupled with an equation describing the build-up and relaxation of the local differential swelling stresses (f) along the plane of the film

$$\delta f/\delta t = (G_o - G_{oo})\delta/s/\delta t + \delta(sG_{oo})/\delta t$$
$$+ ((G_o - G_{oo})^{-1}\delta(G_o - G_{oo})/\delta t - \beta)(f - sG_{oo}). \qquad (2.5)$$

The change in film area is given by

$$A(C_w) = A(0)(1 + k_s C_w). \qquad (2.6)$$

Since the actual area of a thin film is constrained to a uniform value $A(t)$, the creation for local strains is described by the expression

$$s = A(t)/A(x,t) - 1 \qquad (2.7),$$

and corresponding local stresses (f) add up to zero net overall stress.

We justified in [11] that the plasticizing action of solvent and the opposite action of the solute could be described by exponential dependence of the instantaneous and long-term elastic modules on solvent and solute concentrations

$$G_o = G_{oo}exp(-k_{g1}C_w + k_{g2}C_s)$$
$$G_{oo} = G_{ooo}exp(-k_{g1}C_w + k_{g2}C_s). \qquad (2.8)$$

Also, we followed our experimental findings [13] concerning the influence of solvent and solute on relaxation frequency. In particular, the experimental results quoted therein showed that both components cause deviation of β in the same direction, and an exponential dependence is the best description of this effect

$$\beta = \beta_o \exp(k_{d1}C_w + k_{d2}C_s). \qquad (2.9)$$

If the diffusion of solute is not affected by its solubility in polymers we describe its transport by equation

$$\delta C_s/\delta t = (\delta/\delta x)(D_s \delta C_s/\delta x) \qquad 0 < x < 2l \qquad (2.10)$$

where the diffusion coefficient of solute may be assumed to vary with the concentration of sorbed solvent

$$D_s = D_s \exp(-k_{w3}/(C_w + k_{w4})) \qquad (2.11)$$

as postulated by the free volume theory of Yasuda et al. [14].

Equations (2.3)–(2.10) were solved simultaneously following the procedure described in [11] with initial and boundary conditions

$$C_s(x,0) = C_{s0}; \quad C_w(x,0) = 0; \quad \delta C_s(l,t)/\delta x = 0;$$
$$\delta C_w(l,t)/\delta x = 0; \quad C_s(0,t) = 0; \quad C_w(0,t) = C_{w0}.$$

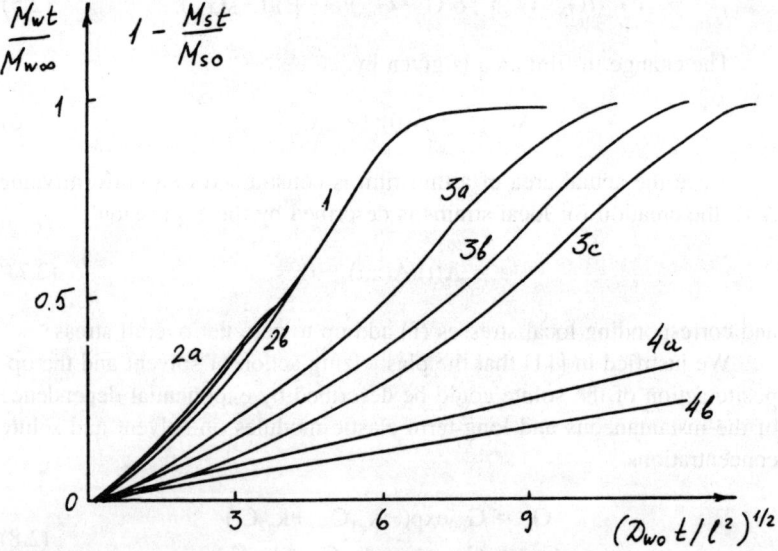

Figure 2.3a. Computed kinetic curves of water uptake (1) and solute desorption (2–4).
1. Water uptake.
2. $D_s(C_{wo})/D_{wo} = 10.$ $D_{so}/D_{wo} = 1000$ (a), 100 (b)
3. $D_s(C_{wo})/D_{wo} = 1.$ $D_{so}/D_{wo} = 1000$ (a), 100 (b), 10 (c).
4. $D_s(C_{wo})/D_{wo} = 0.1$ $D_{so}/D_{wo} = 100$ (a), 10 (b).

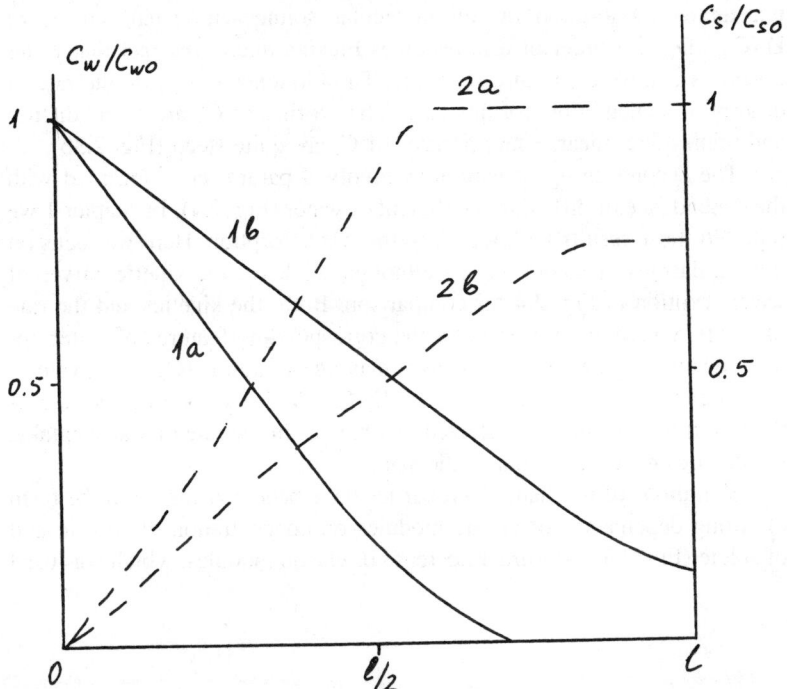

Figure 2.3b. Distribution of solvent (1) and solute (2) in polymer at $(D_{wo}t/l^2)^{1/2}$ = 8 (a) and 12 (b) in case of 2a (Fig. 2.3a).

Following the model and experimental findings [4, 11, 15], the solute affects the kinetics of water sorption in swelling polymers through concentration dependence of elastic modules and relaxation frequency.

Variations of the parameters concerned with the value of D_s and its dependence on the concentration of water provides the opportunity for instructive comparison of the kinetics of water uptake and desorption of solutes of different mobility.

The variations involved the change of D_{so}/D_{wo} in the range from 10 to 1,000, and of k_{w3}/C_{wo} in the range from 3.45 to 13.8, which permitted investigation of the behavior of the system for $0.1 < D_s(C_{wo})/D_{wo} < 10$ and for $0.001 < D_s(C_{wo})/D_{so} < 0.07$.

For all parameter variations indicated above their influence on water uptake is negligible, and the kinetic curve of water sorption is slightly S-shaped (Fig. 2.3a). For high values of $D_s(C_{wo})/D_{wo}$ (of order of 10) the result for solute desorption to follow water uptake has been obtained. Moderate values of $D_s(C_{wo})/D_{wo}$ (of order of 1) lead to non-Fickian release kinetics, the rate of which increases with k_{w3}. The kinetic regularities remain unchanged. Desorption of low-molecular solute with small values of $D_s(C_{wo})/D_{wo}$ (of order of 0.1) becomes Fickian due to the fact that water uptake is completed at an early stage. The influence of k_{w3} on the rate of desorption is negligible for this case. The profiles of C_w are found diffuse and seem quite linear, while profiles of C_s are quite steep (Fig. 2.3b).

The second group of variations involved parameters concerned with the dependence of diffusion coefficient of water (Fig. 2.4). In chapter 1 we reported those results with regard to the water sorption. Here we focus on the regularities of solute release although we leave the kinetic curves of water sorption in Fig. 2.4 for comparison. Both, the kinetics and the rate of solute release are governed by the corresponding features of water uptake (whatever the underlying causes may be) at high $D_s(C_{wo})/D_{wo}$ values.

In contrast to this, for moderate $D_s(C_{wo})/D_{wo}$ values only the rate of the desorption of solute is affected by changes in the rate of water uptake, the kinetics of release remains the same.

Variations of mechanical properties have been considered in the form of strong dependence of elastic modules on concentration of solvent and of solute (Fig. 2.5). A 5-order decrease of elastic modules, which was used

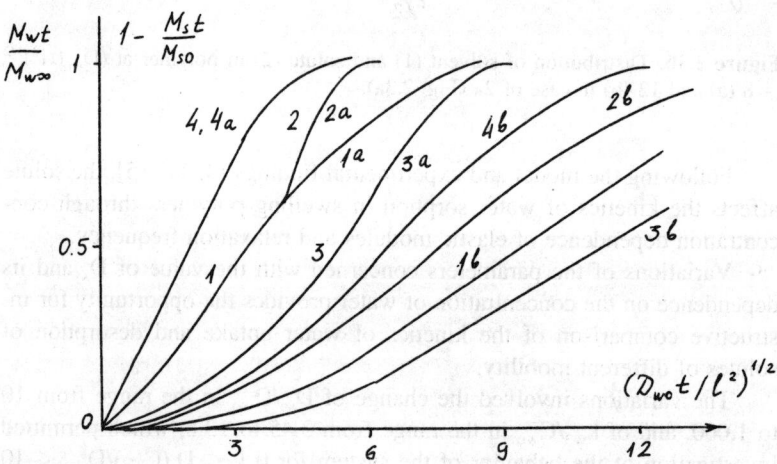

Figure 2.4. Computed kinetic curves of water uptake (1–4) and solute desorption (labeled by letters).
$D_s(C_{wo})/D_{wo} = 10$ (a), 1 (b).
1. $D_w(C_{wo},0)/D_{wo} = e$; $D_w(0, f_{max})/D_{wo} = \exp(f_{max})$; $b_o = 1$
2. $D_w(C_{wo},0)/D_{wo} = 20000$; $D_w(0, f_{max})/D_{wo} = \exp(10f_{max})$; $b_o = 1$
3. $D_w(C_{wo},0)/D_{wo} = e$; $D_w(0, f_{max})/D_{wo} = \exp(10f_{max})$; $b_o = 1$
4. $D_w(C_{wo},0)/D_{wo} = e$; $D_w(0, f_{max})/D_{wo} = \exp(10f_{max})$; $b_o = 100$

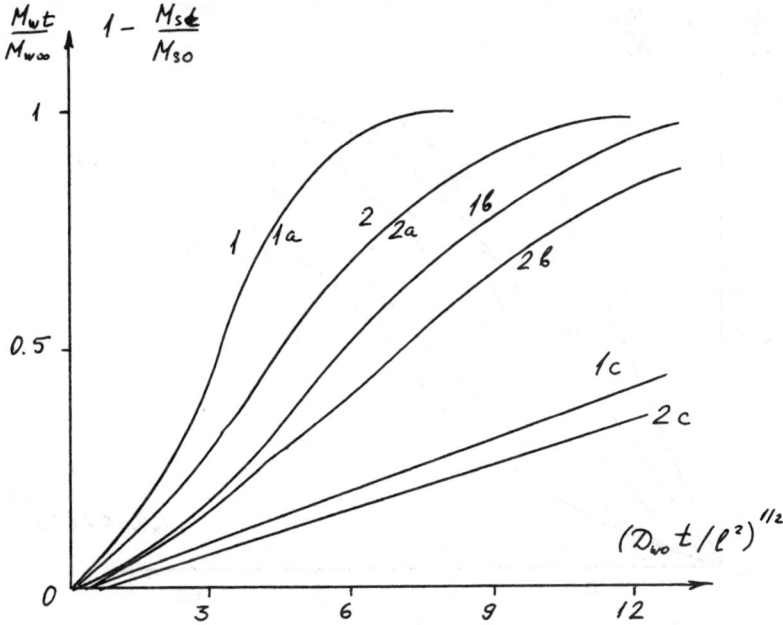

Figure 2.5. Computed kinetic curves of water uptake (1–2) and solute desorption (labeled by letters).
$D_s(C_{wo})/D_{wo}$ = 10 (a), 1 (b), 0.1 (c).
1. $G_i(0, C_{so})/G(C_{wo}, 0)$ = 100,000;   2. $G_i(0, C_{so})/G(C_{wo}, 0)$ = 20

to represent strong dependence on concentration of water, causes almost Case-II diffusion of water until M_{wt}/M_{oo} = 0.6, while nonlinear uptake was observed at higher values of M_{wt}/M_{woo}. The strong dependence of elastic modules on solute concentration is characterized by Fickian water uptake.

The results relating to the kinetics of solute release confirm previous conclusions that the value of $D_s(C_{wo})/D_{wo}$ gives rise to three possibilities of release kinetics, which are (i) desorption closely corresponding to water uptake; (ii) non-Fickian desorption—the rate of which tends to increase as sorption of water accelerates; (iii) Fickian desorption following complete water uptake.

The kinetics of water sorption which correspond to a moderate value of β_0 and weak dependence of β on concentrations of water and solute look Fickian (Fig. 2.6). Because of solute desorption, deceleration of water uptake was observed for the case of strong dependence of stress relaxation on concentration of solute in comparison with the case of the same initial value of β corresponding to unfilled dry polymer. Strong dependence of β

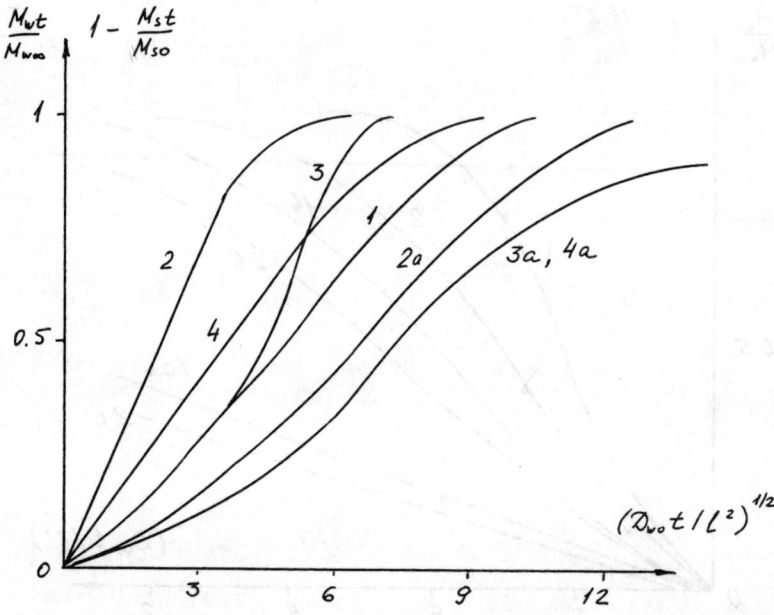

Figure 2.6. Computed kinetic curves of water uptake (1–4) and solute desorption (labeled by a) .

$D_s(C_{wo})/D_{wo} = 1$

1. $\beta_o = 1$; $\beta(0,0)/\beta(C_{wo},0) = \exp(-1)$; $\beta(0, C_{so})/\beta(C_{wo}, 0) = e$
2. $\beta_o = 100$; $\beta(0,0)/\beta(C_{wo},0) = \exp(-1)$; $\beta(0, C_{so})/\beta(C_{wo}, 0) = e$
3. $\beta_o = 1$; $\beta(0,0)/\beta(C_{wo},0) = 0.01$; $\beta(0, C_{so})/\beta(C_{wo}, 0) = e$
4. $\beta_o = 1$; $\beta(0,0)/\beta(C_{wo},0) = \exp(-1)$; $\beta(0, C_{so})/\beta(C_{wo}, 0) = 100$

on concentration of solvent causes further deceleration of water uptake at lower values of M_{wt}. At higher values of M_{wt}, deviation of the kinetic curve of water sorption above the Fickian one was observed, and this leads to higher velocity of water uptake in comparison with the case of strong dependence on content of solute.

Examples of area expansion in relation to overall water uptake (M_{wt}/M_{woo}) are given in Fig. 2.7. The conclusions drawn in [9, 10] remain valid in the presence of the second component. Because the dependence of area expansion is essential for many reasons, further consideration is required to describe possible changes in curves.

The similar approach, which takes proper account of the simultaneous uptake of a liquid leachant by a polymer matrix and the consequent release of a bioactive or other solute incorporated therein, was applied by Petropoulos, Papadokostaki, and Amarantos [16] to predict the kinetics of release of several leachants under various conditions. They showed that the above model can be easily modified or extended, in accordance with the information available about any particular system. It is, therefore, expected that the model will prove useful as a basis for the design of monolithic controlled-release devices of this type or for the evaluation of the leachability of low-level and medium-level radioactive wastes "immobilized" in polymeric matrices.

Penetration kinetics of two other liquids in cellulose acetate were presented by Sanopoulou and Petropoulos in [17]. The viscoelastic response of the polymer was varied by studying transport in uniaxially oriented films with penetration parallel, perpendicular or at a 45° angle to the axis of macromolecular orientation. Rates of penetration are supplemented by information on penetrant concentration and stress distribution within the polymer matrix, deduced from refractive index and birefringence profiles. An example of a successful and physically meaningful interpretation of some

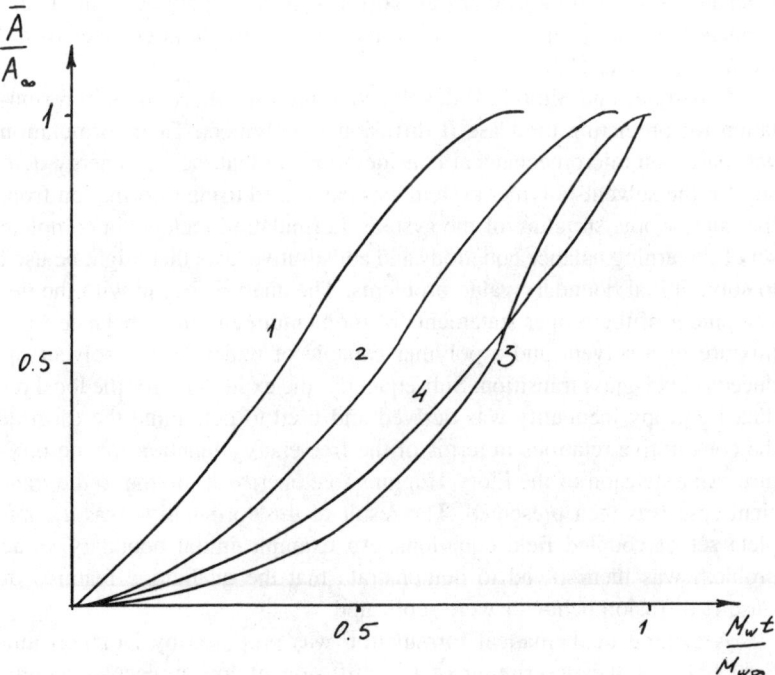

Figure 2.7. Computed correlation of area expansion with corresponding curves of water uptake. Lines labeled: 1 - as in Fig. 2.3a (curve 1). 2 - as in Fig. 2.4 (curve 2). 3 - as in Fig. 2.5 (curve 1). 4 - as in Fig. 2.6 (curve 4).

important features of the observed kinetic behavior were derived by application of the above differential swelling stress model. The study of micromolecular transport in cellulose acetate film reported was extended in [18] by investigation of the effect of different degrees of semipermanent uniaxial macromolecular orientation, produced by prestretching the film under suitable conditions. The changes noted in penetration rate and kinetics and in birefringence, tracer microdensitometry, and microinterferometry profiles for penetration along and across the axis of preferred orientation were reported in detail and their implications in terms of the transport mechanism were discussed.

The longitudinal penetration of micromolecular liquid solvents (acetone, dioxane) or swelling agents (methylene chloride, methanol) into unoriented (unstretched) cellulose acetate film were studied in detail by a variety of techniques, including observation of visible penetrant fronts, birefringence profiles, colored tracer microdensitometry, and microinterferometry. The results provided a fuller picture of the relevant phenomenology than was previously available, leading to further insight into the mechanism of micromolecular transport in stiff-chain polymers and its dependence on the nature of the penetrant and the structural changes of the swelling polymer [19].

Govindjee and Simo [20] developed a microscopic continuum formulation for predicting the Case II diffusion in polymers. Their formulation was based on micromechanical considerations in that the free energy density for the solvent-polymer system was calculated using information from the microscopic structure of the system. Formulation included a complete set of governing balance equations and constitutive laws that might be used to solve initial boundary value problems. The analysis began with the development of the proper statements of momentum and mass balance for a mixture of a solvent and a polymer capable of undergoing a solvent induced rubber-glass transition. Subsequently, the expression for the local reduced entropy inequality was derived and used to determine the form of the constitutive relations in terms of the free energy function for the mixture. An extension of the Flory-Huggins free energy of mixing to the transient case was then presented. The result of this formulation was a complete set of coupled field equations. An example initial boundary value problem was then solved to demonstrate that the qualitative features of Case II diffusion behavior were replicated.

A general mathematical formulation was proposed by Doghieri and Sarti [21] for the description of the diffusion of low molecular weight species into a polymer matrix, in the presence of significant stress and deformation fields. The model was based on the usual momentum and mass balance equations; a crucial point was that it makes use of constitutive equations for the stress and for the diffusive mass flux which are not chosen independently of each other since they must obey precise thermody-namic restrictions; the latter are found in rather general terms. Due to the strong coupling between the mechanical and the diffusive problems, the transfer rate was in general non-Fickian and was affected by the deformation gradient field, by the deformation at the external interface as well as by the mechanical boundary conditions which are exerted. The presence of a stress field resulted in some non-locality effects. The model was explicitly solved for the case of a nonlinear crosslinked network, described by the Flory Erman equation for the Helmholtz free energy.

Hayes and Cohen also proposed a model for non-Fickian diffusion of penetrants into polymers and applied it to study a drug-delivery problem. The model modifies Fick's diffusion equation by the addition of stress-induced flux. A stress evolution equation incorporating aspects of the Maxwell and Kelvin-Voigt viscoelastic stress models completes the model. The relaxation time in the polymer was taken as a function of the penetrant concentration. The system was studied under the assumption that the diffusivity is large. Singular perturbation techniques were used to show that the concentration and stress evolve diffusively for small time, but exhibit steep fronts in a narrow region within the domain for larger time. These predictions were verified numerically for specified parameter values. Finally, the equations were studied in the steady state and were found to predict the evolution of shocks [22].

A model for sharp fronts in glassy polymers was derived and analyzed in [23]. The major effect of a diffusing penetrant on the polymer entanglement network was taken to be the inducement of a differential viscoelastic stress. This couples diffusive and mechanical processes through a viscoelastic response where the strain depends upon the amount of penetrant present. Analytically, the major effect is to produce explicit delay terms via a relaxation parameter. This accounts for the fundamental difference between a polymer in its rubbery state and the polymer in its glassy state, namely, the finite relaxation time in the glassy state due to slow response to changing conditions. Both numerical and analytical perturbation studies of a boundary value problem for a dry glass polymer exposed to a penetrant solvent were completed. Concentration profiles in good agreement with observations were obtained.

The effects of stress whitening on the moisture diffusion rate and concentration in a polymer adhesive containing a secondary phase were investigated in [24]. This was accomplished by performing an absorption test on both stress whitened and virgin samples of the bulk adhesive and comparing the rate and amount of moisture diffusion in each. The presence of stress whitening in samples was not only observed visually, but also confirmed analytically using the "Bilinear RAMOD-2" equation. Experimental results revealed that visibly-present stress whitening resulting from fracture does indeed affect the rate and amount of moisture absorption in a polymer adhesive. Consequently, a diffusion model representing two different regions, stress-whitened and non-stress-whitened, was proposed for path of diffusion in polymer adhesives [24].

Water and ion transport in thin sheets of initially dry, ionic, hydrophilic crosslinked polymers also may be modeled throughout the dynamic swelling process. Hariharan and Peppas expressed the water transport in terms of a non-Fickian equation with a diffusion term containing a Fujita-type concentration-dependent diffusion coefficient coupled with a pseudo-convective term arising from the reasonable assumption that the stress in ionic polymers is proportional to the total number of ionized pendant groups in the polymer. Ion transport was expressed in terms of generalized Fickian equations with water concentration-dependent diffusion coefficients. These equations were solved with appropriate boundary conditions to establish the water uptake as a function of time, pH, and ionic strength in a citrate-phosphate-borate buffer solution. A new dimensionless number, the Stress Swelling Number, λ, was defined to quantify the relative importance of stress in the overall swelling process. Water uptake was a strong function of λ [25]. Water transport into and solute release from ionic, hydrophilic, initially glassy polymeric networks were investigated by deriving a model which takes in consideration the strong concentration dependence of the penetrant and solute diffusion coefficients. The model was solved with appropriate boundary conditions in an ionic buffered solution exhibiting simultaneous ionization of its components. Water uptake and solute release were described in terms of the Swelling Interface Number and other parameters of the polymer/buffer system [26].

Complex mechanical properties of dodecane/polystyrene systems were studied by Kim, Caruthers, and Peppas [27]. The temperature and frequency dependence of the shear modulus were studied for various samples containing different amounts of dodecane. Addition of dodecane lowered the temperature at which mechanical transitions occurred and broadened the transition region. The module decreased with increasing dodecane content. The temperature and concentration dependencies of the shift factor were determined by time-temperature and time-concentration superposition and analyzed by a free-volume model and an entropy model. The concentration effect on the temperature-dependent shift factor was validated by an entropy model. The frequency-dependent complex modules master curves experimentally determined were converted to a discrete relaxation time spectrum; subsequently, the time-dependent material function could be determined from the relaxation spectrum. The relaxation spectrum shifted significantly to shorter times with increasing concentration of dodecane [27].

Mathematical models which assume the concentration and stress-dependent diffusion coefficients were applied to describe moisture transport in epoxy composites using concentration and stress dependent diffusion coefficients. These models take into consideration the stresses induced due to moisture and their influence on the transport mechanism. All governing differential equations are solved with appropriate boundary conditions to obtain concentration and stress profiles. It is concluded that non-Fickian and anomalous transport of moisture in epoxy composites can occur under certain conditions, leading to stress accumulation close to the fibers of the composite [28].

Swelling equilibrium and kinetics of swelling for hyaluronic acid (HA) benzyl ester membranes were examined in media of different pH and ionic strength, and for varying percent esterification of the polymers. The equilibrium swelling properties for polymers with an intermediate (75–85%) degree of esterification in different media can be explained by the Donnan equilibrium theory. However, this theory does not apply to the swelling behaviors of membranes with high ($>$ 85%) and low ($<$ 75%) degrees of esterification, implying that nonideal conditions must also be considered. The kinetics of swelling in media of different ionic strength can be described by a second-order swelling model, with both diffusion and stress relaxation contributing to the process. Ion exchange influences the swelling kinetics of HA ester membranes at lower pH in the medium [29].

Thus, the variation of the mechanical properties of swelling matrix is one of the most important contributing factors in diffusional behavior of water and solute in hydrophilic polymers.

2.1.2. The effect of structure

Two parallel structural processes follow water sorption in hydrophilic polymers as it was shown for films of polyvinylalcohol (PVA) [30]. The first is a relatively fast (seconds) process of moving the polymer chains apart by the water molecules penetrating between them, and a slower process (minutes) of the transition of the paracrystalline structure into the crystalline structure. It was shown that water sorption causes the development of heterogeneity of PVA structure at the level of colloid size, and this is a reversible process. For hydrophilic polymers at temperatures equivalent to T_g the Deborah number is of order of 1. Since the structural rearrangement is reversible we can describe the transport of water and solutes by Eqs. (2.1) and (2.2) with the boundary condition for water

$$C_w (0,t) = C_{w0} + (C_{w0} - C_{woo})[1 - \exp (-t/\tau)]$$

where C_{w0} and C_{woo} are initial and final surface concentration of water.

Table 2.1. Parameters of nonstationary diffusion in polyvinylalcohol and Nylon-6.

Solute–Polymer	$\tau/l^2 \times 10^8$ cm²/s	$D_w \times 10^8$ cm²/s
HCOOH–PVA	0.014	200
CH₃COOH–PVA	0.059	180
C₂H₅COOH–PVA	0.1	170
HCOOH–Nylon	0.06	1.73
CH₃COOH–Nylon	0.51	0.67
C₂H₅COOH–Nylon	0.43	.40

The solute also can affect the structure of polymer, and corresponding relaxation time may be found from sorption experiments from aqueous solutions. The example of the interrelation between transport and structural characteristics of PVA and Nylon-6 are presented in Table 2.1.

Dense PVA membranes of different crystallinity were prepared and studied in pervaporation of water-ethanol mixtures. High selectivity to water was obtained with all types of membranes. Permeation fluxes increase exponentially with the water content in the liquid mixture. At a given water content, the membrane permeability decreases drastically when its crystallinity increases. When the pervaporation temperature increases, the permeation flux increases according to the Arrhenius law, with a permeation activation energy that depends strongly on the crystallinity of the membrane. A permeation model, in which the volume fraction of amorphous polymer intervenes in both the sorption and the diffusion laws, was proposed and validated by the experimental data. With the obtained values of the parameters, the permeation flux can be calculated for membranes of a given crystallinity at any temperature and composition of the water-alcohol mixture [31].

The mechanisms which control the release of dispersed water-soluble drugs from an initially dry hydrogel are complex. The release profile derives from a combination of several contributing factors which may change with time at different rates. It has been possible to isolate controlling factors and investigate their individual contributions to the release kinetics. The hydrogels presented in [32] owe their hydrophilicity to their poly(ethylene oxide) content. They swell and can absorb up to three times their dry weight in water. Having a glass transition temperature (T_g) below body temperature, they are essentially different to those studied theoretically or experimentally, by other groups, which have T_g values above body temperature and are initially glassy. A range of diffusates was studied ranging from low water-soluble prostaglandin E2 to highly water-soluble lithium chloride. Device geometry was restricted to approximations to infinite slabs with more than 85% total surface area over the top and bottom surfaces so that release was predominantly one-dimensional and the controlling variable was thickness. The increase in surface area with time, drug-solubility in the water-swelling matrix, and the presence of crystallinity were shown to be important factors governing the profile and level of release rate with time. It was observed that the release profile could be separated into three parts, the most important being the middle section from early in the release until at least the half-life time. This period could be characterized by the exponential time function, t^n. The diffusional exponent, n, is an important indicator of the release mechanism and ranged from 0.79 to 1, i.e., good anomalous to zero order. This is a highly desirable range of values for controlled release devices. The value of n decreases at late-time. The very early-time release can also show a burst or lag effect depending on the diffusate solubility and its loading in the xerogel [32].

Hydrogels of polyvinylpyrrolidone (PVP) were prepared by using polymer aqueous solutions at 10% w/w and gamma radiation. The swelling experiments were carried out in water at 30°C and followed by weighing. The concentration of effective chains (V_e) and the number-average molecular weight between crosslinks were calculated by considering the chain-end effects. Diffusion and partition coefficients of rhodamine dye were calculated by following the diffusion of the solute out of PVP gel cylinders into water at 30°C. The effect of gel structure on diffusion was studied by preparing gels with various radiation doses and PVP molecular weights. V_e increased with rise in radiation dose, indicating clearly that hydrogels of controlled structure can be prepared. The molecular weight of the polymer altered the gel structure, but these effects were more pronounced at lower absorbed radiation doses. Diffusion coefficient decreased with increase in radiation dose, showing the influence of gel structure on solute diffusion [33].

A bending-beam technique was used to in situ monitor the diffusion of water in various polyimide films. The polyimides studied were pyromellitic dianhydride-4,4'-oxydianiline (PMDA-ODA), pyromellitic dianhydride-p-phenylenediamine (PMDA-PDA), and 3,3',4,4'-benzophenone tetracarboxylic dianhydride-p-phenylenediamine (BPDA-PDA), and their blends and random copolymers. The diffusion of water in these films obeys Fick's law. In PMDA-ODA, the mean diffusion constant was 5.2×10^{-9} cm^2/s, for thickness ranging from 6.7 to 27.3 μ. In PMDA-PDA, it was 2.0×10^{-9} cm^2/s for thickness ranging from 7.3 to 20.0 μ, and in

BPDA-PDA, 0.27×10^{-9} cm^{-2}/s for thickness ranging from 4.8 to 21.0 μ. In the blends and random copolymer with 50% wt PMDA-ODA and 50% wt PMDA-PDA, the diffusion constants were slightly smaller than those in the pure PMDA-ODA, but much larger than in the pure PMDA-PDA. On the contrary, in those with 50% wt BPDA-PDA and 50% wt PMDA-PDA, the diffusion constants were much smaller than those in the pure PMDA-PDA, but slightly larger than in the pure BPDA-PDA. These diffusion constants were primarily affected by the chemical structure of the imide molecule. The morphology, such as crystallinity, of the films has played a secondary factor. Hygroscopic stresses due to water uptake in all the studied films increase as the film thickness increases. It can be attributed to the fact that the film orientation decreases with the increase of thickness [34].

The influence of molecular primary and secondary structure on the transport of water has been investigated for a series of chemically similar polyimides using an infrared ATR method. The diffusivity of water generally decreased with increased chain backbone stiffness (as determined from the mechanical beta-transition), and the activation energies for diffusion generally increased with backbone stiffness. For a given backbone composition, the diffusivity was also shown to markedly increase as the density of the amorphous phase decreased. The vibrational spectra show that the absorbed water partitions into a bimodal distribution of sites, presumably representing single and clustered water molecules. This distribution changed markedly with saturation and was hence a sensitive probe of the local heterogeneity of the glassy structure [35].

The release of drugs, peptides, and proteins from hydrogels is controlled by the macromolecular structure of the carrier as defined by the degree of crosslinking, the degree of swelling, and related parameters. Ionic interactions play an important role in drug transport. For swelling-controlled systems, the coupling of diffusion and macromolecular relaxation control the mechanism of drug release, providing conditions of zero-order release. Application of such systems in various delivery situations was discussed in [36] with emphasis on the delivery rate and the stability of the releasing bioactive agents.

M. McNeill and Graham [37] examined the state of water-association with poly(ethyleneoxide), as evidenced by diffusivity, in a series of crosslinked polyurethanes made from poly(ethylene glycols) of a range of molecular weights. As a subsidiary underpinning exercise, the correlation of diffusivity with water content at relatively high levels of swelling (>45%) using a variety of semi-empirical equations was analyzed. Three water-soluble compounds with similar molecular weights and which exhibit minimal interaction with the polymer, as shown by their partition coefficients, were chosen. These were proxyphylline, morphine hydrochloride, and caffeine. The best statistical correlation of the data was obtained for plots of: (a) diffusivity against weight percent water; and (b) log diffusivity against the reciprocal of the weight percent of water in the hydrogels. Proxyphylline results for the high levels of swelling compositions were augmented with data from lower swelling compositions and a clear break in the slope of diffusivity against percentage of water in the swollen hydrogel was obtained. This indicated a change in the nature of the diffusion at this point. The probability of this transition point corresponding to a change for diffusion through water bound as trihydrate to diffusion in free water was discussed [37].

Diffusion of uniformly dispersed tritium-labeled estradiol in water-swollen poly(ethylene oxide)-based hydrogels was studied by assaying the release of solute from cylindrical hydrogels into a finite volume of solution at 37°C. Fractional release followed a $t^{0.5}$ relationship for a range of radii between 0.20 and 0.35 cm and different polymer compositions with equilibrium water uptake of 220–750 parts per hundred dry polymer. Values of the diffusion coefficient were calculated from the fully swollen hydrogels, which represent the swelling and release profile quite accurately [38].

Diffusion of dyes into fibers does not accurately follow usual laws for diffusion into liquids or amorphous polymers. Apparently, the reason for this is that the fraction of a fiber which has sufficient molecular mobility to allow diffusion is temperature dependent. Thus, an understanding of diffusion needs to deal with both thermal mobility and the effects of various constraints, on thermal motion of noncrystalline chain segments. A method for converting fiber dynamic mechanical data into two parameters, an internal viscosity (α) and a mobile noncrystalline fraction (X), the ratio of which behaves very much like dye diffusion in nylon and PET fibers, are reviewed. The effects of water, polymer, and structure on α and X also are discussed in [39].

The problem of transport with possible solute binding and consumption was considered for membranes in series, as well as other multilaminar media, in which transport and reaction processes are linear and time invariant. By analogy with electrical networks, the properties of a membrane were summarized by an admittance (or alternatively, impedance) matrix that relates fluxes to concentrations on either side of the membrane. The elements of the admittance matrix were shown to be related to the permeability, solute consumption, and certain lead and lag-time parameters of the membrane. The admittance matrix can then be transformed into a transmission matrix. The product of transmission matrices for two membranes in series was equal to the transmission matrix for the series combination. From these facts, simple algebraic relations were derived that relate parameters for membranes in series to the same parameters for the individual membranes, where the latter parameters can be estimated theoretically or regarded as phenomenological quantities to be determined by experiment. These relations have differential counterparts when the properties of the medium vary continuously in space. In the absence of solute consumption, integral relations were also derived for continuously varying media. Results for continuous media can be used to extend the analysis to cases where convection and/or an externally applied force field is present. The relationship of the lead and lag-times to mean first passage times is noted in [40].

The coprecipitates were prepared by a solvent technique using Eudragit E as carrier and indomethacin as a model drug. X-ray diffractometry, differential scanning calorimetry (DSC), and wettability tests were employed to investigate the physical state of the studied formulations. Up to 50% of indomethacin can be dispersed in an amorphous state in Eudragit E. The influence of the pH on the in vitro release of solid dispersions was evaluated. Because of the good solubility of Eudragit E at pH 1.2 a fast dissolution rate of the drug was observed while a marked delay was noticed at pH 7.5 where the polymer is only permeable to water. At pH 5.8 the kinetics of drug release can be modulated by the drug/polymer ratio. The dissolution rate of the drug can be increased by decreasing its amount in the coevaporate [41].

Controlled release tablets of indomethacin have been prepared, using Eudragit RS as a matrix. Preparation was by compression, after either mixing or granulating the ingredients. The content of polymer was found to influence the drug-release rate during in vitro dissolution tests. Granulated products showed significantly slower drug release than directly compressed products containing the same ingredients. Measurements of the wettability of the tablets, as determined by preparing compressed plates and using a Wilhelmy technique, when plotted as a function of release rate, resulted generally in an inverse proportionality. This was explained as being due to the wettability relating to the effective surface area of the product. The mechanism of drug release was assessed by calculating the thermodynamic parameters of activation from apparent release rate constants obtained at four different temperatures. Compensation analysis was undertaken, in the enthalpy-free energy domain, to investigate the presence of common mechanisms of drug release. The change in process from direct compression to granulation did not alter the release mechanism [42].

The diffusional behavior of sodium salicylate (SSA) from a new chemically crosslinked copolymeric hydrogel of 2-hydroxyethyl methacrylate with N,N'-dimethyl-N-methacryloyloxyethyl-N-(3-sulfopropyl)ammonium betaine was reported. The drug release into water was effected under sink conditions and the kinetics at 310 K were followed by continuous measurement of conductivity. A modified form of Fick's second law has been derived to take into account the dimensional changes of copolymer during drug release, which were measured photographically. The procedure was used to evaluate the diffusion coefficient D_s for transport of SSA into water. The value of $D_s = 4.4 \times 10^{-7}$ cm^2/s yielded over a range of drug loading is smaller than the load-dependent values obtained when allowance is not made for dimensional changes. The procedure was also used to determine the diffusion coefficient for water into polymer [43].

Drug-release mechanisms of swellable systems for controlled drug administration were investigated in [44]. The variations of the matrix relaxation and drug diffusion rates were quantified by measuring the surface area exposed during matrix swelling and drug release as a function of impermeable coating coverage and location. Four different types of matrices, partially coated on various sides, were investigated in order to elucidate the role of the swelling behavior on the release from such delivery systems, especially in view of the three-dimensional nature of the swelling phenomenon. Dependence of the release kinetics on the matrix surface area was assessed. A new dimensionless number, the swelling area number S_a was defined for evaluating the significance of the relative rate of matrix swelling variation and drug diffusivity. The systems studied were produced by partial coverage of the release area of tablets by an impermeable coating [44].

Graft copolymers of poly(ethylene glycol) with poly(methacrylic acid) were prepared by reaction of poly(ethylene glycol) methacrylate macromonomer with methacrylic acid in the presence of tetraethylene glycol dimethacrylate as a crosslinking agent. The materials were swollen in aqueous solutions with the pH ranging from 1.5 to 12.0. Dynamic swelling studies were performed. The swelling rates were lowest in copolymer networks with 1:1 ratios of ethylene oxide to methacrylic acid. These results were attributed to the combined effects of diffusion and complex dissociation during swelling. Solute release studies were carried out to investigate the effect of pH and swelling on the release behavior. Lowest solute release rates were observed in complex-forming networks, verifying the relationship between complexation, swelling, and solute permeability [45].

2.1.3. Effect of the solubility of solute

The application of low-soluble compounds in systems of controlled release requires accurate consideration of the effect of solubility on the kinetics and rate of release. For this reason, we extended the model described by Eqs. (2.3)–(2.11) to conditions where the solute is not completely soluble in polymer [46]. This approach was suggested by Petropoulos et al. [47] and found another application reported in [16].

Thus, here we describe transport of solute by equation

$$\delta C_s/\delta t = (\delta/\delta x)(D_s \, S_s \, \delta a_s/\delta x) \qquad 0 < x < 2l \qquad (2.12)$$

In equation (2.12) a_s is the activity of the solute in the matrix defined as equal to that of the same solute in an external phase in equilibrium. Thus the relation between C_s and a_s is given by the equilibrium sorption (or partition) isotherm, which governs the sorption of the solute in question by the solid matrix from the said external phase. Therefore, S_s and D_s are, respectively, the thermodynamic solubility and diffusion coefficients.

For the solute we have

$$S_s = C_{ss}/a_s$$

where C_{ss} is the concentration of dissolved part of solute.

The activity, neglecting nonideality effects, may be represented by $a_s = c_{ss}/c_{ss}^\circ$, where c_{ss} denotes concentration of the solute in aqueous solution and c_{ss}° the corresponding saturation value. The sorption isotherm relating C_{ss} to c_{ss}° may be written as

$$C_{ss}^\circ = k_{ss} C_w V_w c_{ss}. \qquad (2.13)$$

It is often assumed for swelling polymers that $k_{ss} = 1$; the implication being that C_{ss} corresponds to the amount of the embedded solute that can be dissolved by the imbibed water, assuming the solvent power of the latter to be the same as that of free water. This assumption is often satisfactory for lower amount of solute and high C_w, but can be expected to be progressively lost with C_s increase or c_{ss}° decrease. From S_s definition and Eq. (2.13) we obtain

$$S_s = k_{ss} C_w V_w \, c_{ss}^\circ = K_s \, C_w. \qquad (2.14)$$

Thus for any C_w, S_s represents the saturation value of C_{ss}, and, at any point within the matrix during the diffusion process,

$$C_{ss} = S_s, \quad a_s = 1 \qquad \text{if} \qquad C_s > (\text{or} =)S_s$$
$$C_{ss} = C_s, \quad a_s = S_s/C_{ss} \qquad \text{if} \qquad C_s < S_s.$$

For high values of $D_s(C_{wo})/D_{wo}$ (of order of 10) the solubility of solute affects neither kinetics nor rate of its desorption. Such effect becomes visible for moderate values of $D_s(C_{wo})/D_{wo}$ (of order of 1) although it

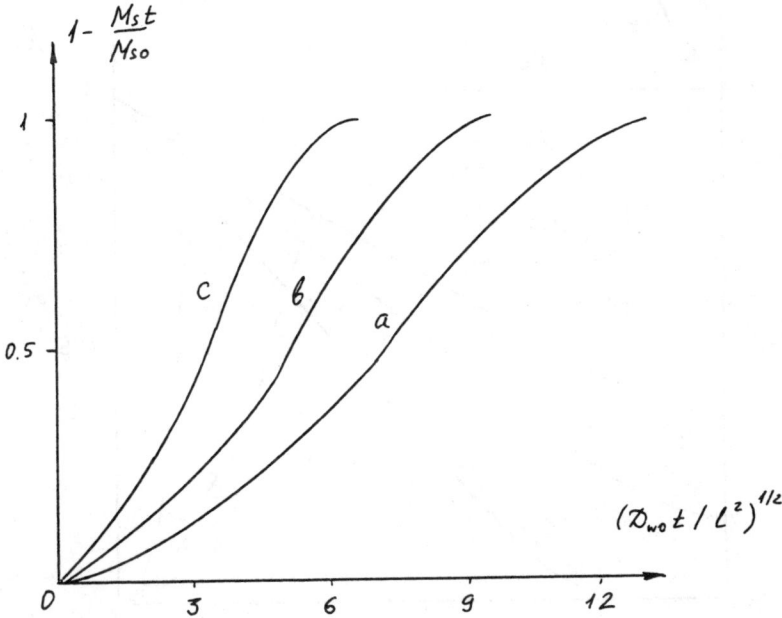

Figure 2.8a. Computed kinetic curves of the release of solutes of different solubility.
$D_s(C_{wo})/D_{wo} = 1$
$K_s = 0.1$ (a), 0.3 (b), 1 (c).

relates only to the rate of desorption (Fig. 2.8a). The kinetic regularities remain unchanged. Such a regularity of solute desorption in swelling polymers is in contrast with the solute desorption from nonswelling matrix, which will be considered below. The profiles of C_w are not affected by solubility and remain linear, while the steep character of C_s profiles develops as solubility of solute reduces (Fig. 2.8b).

The effect of solubility on the kinetics of water sorption is essential for the case of strong dependence of relaxation frequency on the concentration of solute. Fig. 2.9 illustrates how solubility of the second compound alters either the kinetics or the rate of water sorption.

Controlled release systems of theophylline, proxyphylline, and oxprenolol-HCI exhibiting modulated drug delivery were prepared by using pH-sensitive anionic copolymers of 2-hydroxyethyl methacrylate with acrylic acid or methacrylic acid. Drug-release studies were carried out in simulated biological fluids. The initial drug-release rates and the drug-release mechanisms were dependent upon the pH and ionic strength of the

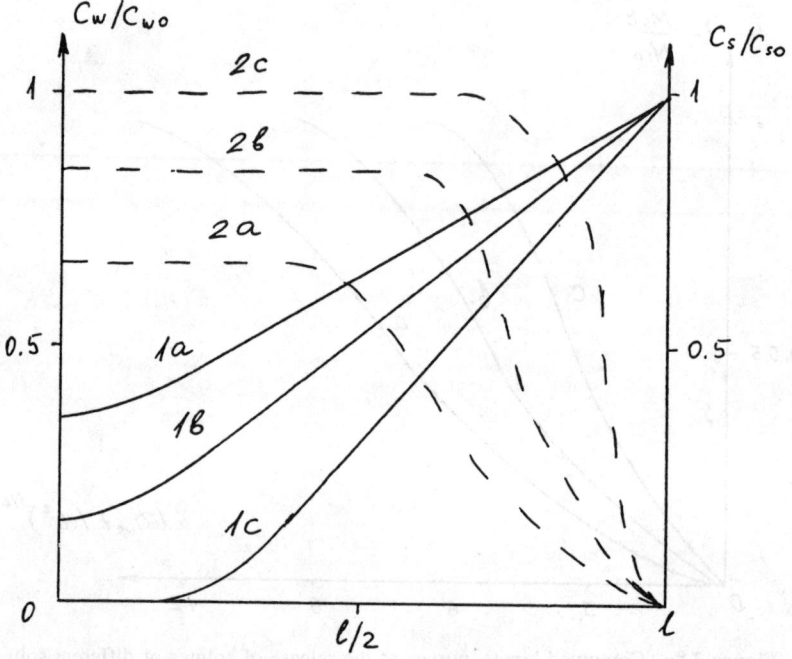

Figure 2.8b. Distribution of solvent (1) and solute (2) in polymer at $(D_{wo}t/l^2)^{1/2}$ = 8 in case of Fig. 2.8a. The labels are the same.

buffer solution as well as its salt composition. Initial drug diffusion coefficients in these swelling-controlled release systems were calculated from the release curves; they were of the order of 10^{-7} cm²/s and were dependent upon the degree of swelling. The drug-release mechanism was non-Fickian in all the dissolution media studied. Lowest release rates were observed for drug release from nonionized polymer networks in agreement with the relationship between ionization, swelling, and drug release [48].

The release properties of calcium alginate minimatrices were studied in media of various compositions. Three drugs with different aqueous solubility (paracetamol, theophylline, and chloramphenicol) were incorporated as model substances and their release rates were investigated in 0.1 M HCl and water. The theophylline release was also studied in simulated gastric fluid (SGF), simulated intestinal fluid (SIF), 0.034 M NaCl, and 0.1 M NaCl. Additionally, the simultaneous liberation of calcium ions from the carrier material into the different media was analyzed and illustrated by means of calcium release curves. Only when pure water was applied as release medium were the matrices able to extend the release of the two least soluble model drugs, theophylline and chloramphenicol. In all other media the drug release proceeded much more rapidly, due to various transformations in the carrier material. The cross-linking calcium ions were rapidly discharged from the matrices in the presence of acid, and the carrier material was converted to alginic acid. Although the transformation did not change the morphology or the swelling behavior of the matrices, it destroyed their ability to provide retarded drug release. In the NaCl solutions and SIF, the calcium ions were partly exchanged by the nongelling sodium ions or sequestered by the phosphate. This caused swelling and, in the latter case, dissolution of the matrices, and induced a rapid release of the encapsulated drug. Due to the pronounced sensitivity towards the composition of the release medium and the rapid drug release in media of physiological relevance, it was concluded that the minimatrices do not seem applicable as an oral controlled release system [49].

An NMR imaging method was developed to estimate the rate of water movement in slow-release capsule matrices of pseudoephedrine HCl

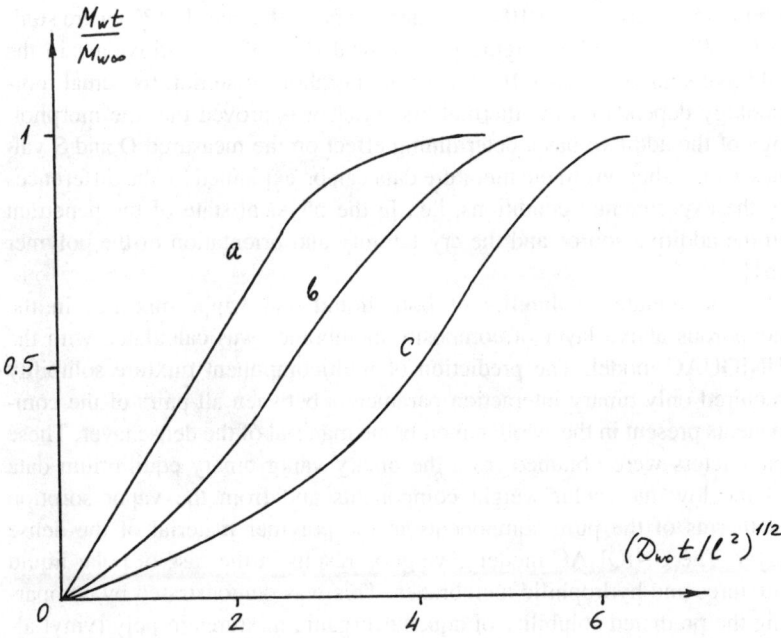

Figure 2.9. Computed kinetic curves of water uptake in presence of solute of different solubility.
$D_s(C_{wo})/D_{wo} = 1$; $\beta_o = 1$; $\beta(0,0)/\beta(C_{wo},0) = \exp(-1)$; $\beta(0, C_{so})/\beta(C_{wo}, 0) = 100$
$K_s = 0.1$ (a), 0.3 (b), 1 (c).

and hydroxypropyl cellulose (HPC). Test capsules were first placed in a USP method 2 (paddles, 50 rpm) dissolution apparatus. Each plug was removed from the dissolution medium at predetermined times, blotted dry, and placed within the magnetic field of a General Electric 400-MHz widebore NMR spectrometer equipped with a microimaging accessory. Images were recorded along the transverse plane of each plug. The water penetration rate was determined by comparison of the cut and weighed contour plots of the images acquired. After 1 hour, the plugs tamped to 200 N exhibited water penetration to the center, while only 45% of the drug was released. The percentage dry matrix was fitted to the Jest equation to obtain a diffusion coefficient of 4.15×10^{-6} cm^2/s. NMR imaging was set forth as an important and practicable technique to investigate drug formulations. In the HPC matrix system of this study, the results of NMR imaging convincingly revealed the rate of hydration front penetration not to be a rate-limiting step in the drug-release process [50].

Transport properties of antioxidants in polymers have a very important role in their effectiveness. Lifetime of a product is strongly influenced by the physical loss of the stabilizer. The diffusion coefficient (D) and solubility (S) of Irganox 1010 in low-density polyethylene (LDPE) were studied at 45 and 80°C, changing the physical state of the antioxidant in the additive source. Irganox 1010 is a polymorphous material; its actual morphology depends on the thermal history. It was proved that the morphology of the additive has a determining effect on the measured D and S values. Contradictions in the literature data can be explained by the differences in the experimental conditions, i.e., in the physical state of the penetrant in the additive source and the crystallinity and orientation of the polymer [51].

The nonideal solubility of both liquid and vapor mixtures in the nonporous active layer of composite membranes was calculated with the UNIQUAC model. The prediction of multicomponent mixture solubility required only binary interaction parameters between all pairs of the components present in the swollen membrane material of the dense layer. These parameters were obtained from the binary vapor-binary equilibrium data of the low molecular weight components and from the vapor sorption isotherms of the pure components in the polymer material of the dense layer. The UNIQUAC model gave good results in the case of polar liquid mixtures and hydrophilic membranes. This was demonstrated by comparing the predicted solubility of aqueous/organic mixtures in poly (vinyl alcohol) membrane material with the experimental data of six mixture systems [52].

Partition coefficients obtained from liquid chromatography were compared to those obtained from equilibrium sorption experiments for various alcohol-water-cellulose acetate systems. In the liquid chromatography experiments cellulose acetate powder was used as the stationary phase with water as the mobile phase and alcohols injected into the mobile phase. In the equilibrium sorption experiments, cellulose acetate power was equilibrated with an alcohol-water solution. The distribution coefficients determined using liquid chromatography and those determined using liquid phase sorption exhibit similar trends with changing solute structure. However, the chromatographic distribution coefficients are, in general, smaller than the equilibrium distribution coefficients. This discrepancy is attributed to the fact that equilibrium is not attained in the chromatographic column. The fact that the distribution coefficients are in reasonable agreement indicates that while the kinetics of diffusion prevent a true equilibrium distribution coefficient from being obtained from the chromatography column, the results of the chromatographic experiments give a qualitative representation of the affinity of the polymer phase for each solute [53].

Particle size usually has an important effect on the rate of drug release from sustained release systems. The nature of that effect depends on the geometry of the system and the mechanism of drug release. Earlier work showed that inferring the mechanism of drug release from ensembles may be misleading. Su et al. [54] reported that ibuprofen release from cetostearyl alcohol matrices with two different particle size ranges, was investigated both experimentally and using Monte Carlo simulation. It was shown that with particles with a mean radius of 431 μ, of the four models used, only the cube-root model gave a satisfactory fit to the experimental data. With the smaller particles with an average radius of 106 μ, both the first-order and the cube-root models were satisfactory. When an equi-weight mixture of the particles was investigated, both the first-order and the cube-root models gave simulated values which predicted the experimental values well. However, the first-order model tended to overestimate drug release in the initial stages ($<$ 30%). The value of the work described is that given the profiles of drug release from particles from two size ranges, it seems possible to predict precisely drug release of mixtures of particles drawn from those two populations. Conversely, it apparently is possible to obtain a specific drug-release profile intermediate between those of the original profiles given those profiles. Such flexibility would enable closer tailoring of drug release from sustained release spherical matrices of the type described without extensive trial and error experimentation [54].

Solid dispersions of carbamazepine with polyvinylpyrrolidone/vinylacetate copolymer at different loading ratios were prepared to study the influence of this copolymer on the solubility and dissolution rate of the drug. The greatest increase in the dissolution rate of carbamazepine was obtained from solid dispersion at the 1/4 w/w drug/polymer ratio. Also the influence of pH on dissolution was studied. The data indicated that the release profile of the drug was not modified by a change in pH. Several kinetic models have been applied in an attempt to describe the mechanism of drug dissolution. Physical characterization of the prepared systems was carried out by differential scanning calorimetry (DSC), X-ray diffractometry, and wettability studies [55].

The influence of the solute solubility grows as the hydrophilicity of the polymer reduces.

2.2. MODERATELY HYDROPHILIC AND HYDROPHOBIC POLYMERS

As pointed out in chapter 1, even for very hydrophobic polymers the diffusion of solutes is, in general, affected by thermodynamics and kinetics of water uptake.

Guo [56] described the effects of water on the properties of ethylcellulose films. The change of density, morphology, porosity, and drug transport properties of ethylcellulose films with the water contents in the polymer solutions were investigated. Due to the strong-hydrogen-bonding property of water, the process of dissolving and mixing of ethylcellulose in a solvent (ethanol) was dramatically changed when a nonsolvent (water) was added in this system. Since the solubility parameter difference between water and the rest of components, and the evaporating rate difference between ethanol and water could cause the phase separation of polymer solution during the film-forming process, the porosity of ethylcellulose films increased with water content in the polymer solutions. The film density and the drug permeability of ethylcellulose films were found to decrease and increase with water content in the polymer solutions, respectively; and these results were very consistent with the porosity increase of ethylcellulose films [56].

One of the main aims in the design of a system for controlled release is to alter the hydrophilic/hydrophobic balance in the polymer. For this reason the principle difference between polymers should be searched not in the equilibrium concentration of water, but in effect of the limited water sorption on kinetics and rate of release. This effect is visible to certain extent for all polymers which sorb less than 10% of water in equilibrium, and even for a dozen highly swelling copolymers and polymer blends with hydrophobic fragments, domains, etc. The load of hydrophilic fillers and additives also does not cancel the influence of the polymer nature. For the aforesaid and many other reasons, we combine the above polymers in one group in order to consider it in this paragraph. Our consideration is based on the comprehensive analysis of these systems made by Petropoulos [47, 16].

The simplified theoretical model for elution of a solute from an initially dry hydrophilic polymer matrix with simultaneous imbibition of water (Eqs. 2.1–2.2) may be applied for less hydrophilic polymers [57, 58]. However, the applicability of the model of Korsmeyer, Peppas, and their colleagues [57, 58] is severely limited by certain simplifications.

First, the equilibrium water uptake of the matrix tends to increase as the amount and/or water solubility of the embedded solute increases [59]. Secondly, the total concentration of solute (C_s) at any point in the matrix is, in general, composed of a mobile or "dissolved" part (concentration C_{ss}), and an immobile or "dispersed" part (concentration $C_s - C_{ss}$). C_{ss} is the increasing function of the water content of the matrix. Thirdly, and largely as a consequence of the foregoing points, the driving force of diffusion cannot be properly formulated in terms of concentration gradients [60]. Accordingly, Petropoulos proceeded to formulate a more general and rigorous treatment, in which proper account is taken of all the above points. He formulated the dimensionless diffusion equation in terms of chemical potential gradient $(\delta\mu/\delta x)$ for both of the diffusion species

$$\delta C_w/\delta t = (\delta/\delta x)(m_{Tw}C_w\delta\mu_w/\delta x)$$

$$\delta C_s/\delta t = (\delta/\delta x)(m_{Ts}C_s\delta\mu_s/\delta x)$$

(2.15)

where m_T is the "thermodynamic" mobility coefficient. Since $\mu = RT\ln a$, Eqs. 2.15 become

$$\delta C_w/\delta t = (\delta/\delta x)(D_{wT}S_w \, \delta a_w/\delta x)$$

$$\delta C_s/\delta t = (\delta/\delta x)(D_sS_s \, \delta a_s/\delta x)$$

(2.16)

where $D_i = m_{Ti}RT$ and $S_i = C_i/a_i$ are thermodynamic diffusion and sorption coefficients. Thermodynamic diffusion coefficient is not identical with the "Fick diffusion coefficient," unless S = constant.

Therefore, Eqs. (2.1)–(2.2) should be replaced by Eqs. (2.16) with boundary conditions

$$a_s(0,t) = 0; \quad a_w(0,t) = 1$$

$$\delta a_s(1,t)/\delta x = 0; \quad \delta a_w(1,t)/\delta x = 0$$

(2.17)

$$a_s(x,0) = a_{s0}; \quad a_w(x,0) = 0$$

In Eqs. (2.17) a_w is given by the relative water vapor pressure. The experimental water sorption isotherms in hydrophobic and moderately hydrophilic polymer matrices indicate that S_w usually tends to increase with

a_w and also with the salt load C_s [47, 61]. Petropoulos et al. suggested to represent this dependence semiquantitatively by the relation

$$S_w = C_w/a_w = (K_{w1} + K_{w2}a_w)(1 + K_{w3}C_s). \qquad (2.18)$$

In the meantime, we found that water solubility in poly(hydroxybutyrate) containing potassium and/or sodium chloride is better described by exponential dependence

$$S_w = K_{w1} \exp(K_{w2}a_w + K_{w3}C_s). \qquad (2.19)$$

Obviously, both expressions may be further modified to represent any other particular system more accurately. Hence, the local a_w value may be found at any point of matrix, during the diffusion process from the local C_w following Eqs. (2.18) or (2.19).

For the solute we get $S_s = C_{ss}/a_s$. As pointed out in section 2.1.3 the sorption isotherm relating C_{ss} to the concentration of solute in aqueous solution may be written in form (2.13). For low hydrophilic polymers one should expect that k_s is progressively less 1 as C_w is reduced. If any experimental information on k_s behavior is available, Eq. 2.13 allows evaluation of C_s for that particular case. Thus we can obtain

$$S_s = C_s/a_s = k_{ss}C_wV_wc^\circ_{ss} = K_sC_{ss}. \qquad (2.20)$$

As in hydrophilic polymers, we can use Eq. 2.20 to represent the saturation value of C_{ss} at any point within the matrix.

Both diffusion coefficients D_w and D_s also must be formulated as functions of C_w in general forms (2.4) and (2.11), respectively. In comparison with hydrophilic polymers we may expect here the weak dependence of the diffusion coefficient of water on swelling stress ($k_{w2} \sim 0$). Although the general dependence of D_w on C_{ws} remains the same, for hydrophobic and moderately hydrophilic polymers the value of k_{w1} may be positive, negative, or close to zero. The concentration dependence of C_w is always assumed to follow the "free volume theory" of Yasuda and Peterlin [14].

Therefore, effect of solute on water sorption in nonswelling polymers is defined, mostly, by the dependence of the solubility of water on the concentration of solute which in addition presents in these polymers either in dissolved or dispersed forms. This general difference between swelling and nonswelling polymers defines further differences in regularities of multicomponent transport.

Following the above theoretical consideration the major factors which affect the kinetics and rate of water sorption and solute release must be

(i) polymer structure, which is responsible for the amount of solute that could be dissolved;

(ii) solubility of components, first of all of solute; and

(iii) presence of third components, such as fillers, additives, stabilizers, impurities, etc., which can change (sometimes, dramatically) the hydrophobic/hydrophilic balance in the matrix.

All these factors are considered below with references to corresponding prediction of the above model.

2.2.1. The effect of polymer structure

Saltzman et al. [62] presented a new method for controlling the rate of antibody (Ab) release from an inert matrix composed of poly(ethylene-co-vinyl acetate) (EVAc), a biocompatible polymer that is frequently used to achieve controlled release. Using supercritical propane, a parent sample of the copolymer of ethylene with vinylacetate (EVAc) (M_n = 70 kDa, M_w/M_n = 2.4) was separated into narrow fractions with a range of molecular weights ($8.7 < M_n < 165$ kDa, $1.4 < M_w/M_n < 1.7$). Solid particles of Ab were dispersed in matrices composed of different polymer fractions, and the rate of Ab release into buffered saline was measured. The rate of Ab release from the EVAc-copolymer matrix depended on molecular weight: more than 90% of the incorporated Ab was released from low molecular weight fractions ($M_n < 40$ kDa) during the first 5 days of release, while less than 10% was released from the high molecular weight fraction ($M_n > 160$ kDa) during 14 days of release. No significant differences in polymer composition, glass-transition temperature, or crystallinity were identified in the different molecular weight fractions of EVAc-copolymer. Mechanical properties of the polymer did depend on the molecular weight distribution, and correlated directly with Ab release rates. Because it permits rapid and reproducible fractionation of polymers, supercritical fluid extraction was used to modify the performance of polymeric biomaterials.

According to the nature of interaction of polyethylene terephthalate (PET) with binary solutions, two groups of systems had been identified. Those belonging to group A were aqueous solutions of acetone and dioxane, as well as acetone solutions of isopropanol. Systems belonging to group B were: acetone-carbon tetrachloride, dioxane-carbon tetrachloride, and dioxane-hexane. They are characterized by a positive deviation of swelling from additivity and a smooth trend of the crystallinity and permeability curves. The treatment of PET films with group A solutions was a simple method of modifying the polymer structure, leading to an increase in its permeability. Preliminary biaxial orientation of PET rendered it suitable for use in the separation of liquids which crystallize an unoriented polymer so strongly that it loses its mechanical strength [63].

Drawn PET fibers (draw ratio = 5:1) were treated unrestrained in dimethylformamide and benzaldehyde at 100°C for varying amounts of time. A rapid diffusion of liquids into the polymer occurred, the latter experiencing considerable retention of the absorbed liquids at 60°C. The increase in chain mobility brought about by both the liquids absorbed and the liquid-induced crystallization that followed allowed orientation strains existing in the polymer structure to be relieved. This brought about a 20% shrinkage in the polymer and a change in its overall geometry. Subsequently, a sharp drop occurred in the level of liquid retention within the polymer signifying expulsion of liquids from the crystallizing domains and ultimately a leveling off (equilibrium) in the weight retention kinetics.

However, precrystallization of the fibers at 200°C for 3 hours using dry hot air prior to liquid treatment study at 100°C and retention at 60°C led to a linear increase in the amount of liquid retained in the polymer as a function of time until an equilibrium was established. The essential feature of the latter liquid retention kinetic is that a plot of the amount of relative liquid retained, that is, M_t/M_{oo} as a function of square root of time ($t^{1/2}$) agrees with Fick's standard diffusion process [64].

Polymer-coated attenuated total reflection (ATR) elements have been used to compare the diffusion behavior and enrichment from aqueous solutions of three different chlorinated hydrocarbons (CHCs)—monochlorobenzene (MCB), chloroform (CF), and tetrachloroethylene (TeCE)—into different polymers. The influence of polymer properties such as glass transition temperature and crystallinity and the effect of the polymer background IR absorption and varying thickness of the polymer membranes to the detectability were investigated. The crystallinity and the glass transition temperature had a very pronounced influence on the velocity of the diffusion process, whereas the partition coefficient influenced the amount of CHC diffusing into the polymer membrane. The time constants for 90% saturation of the polymer with the test analytes were in the range of 8 to 40 minutes, depending on the nature of the polymer and analyte. A linear calibration graph was obtained for simultaneous detection of all three test analytes in the range from 5 mg/L to 100 mg/l CHC with detection limits of 1.5 to 2 mg/l. Coefficients for CHC partitioning between water and polymers measured by ATR/IR were in good agreement with values determined by GC/MS [65].

The diffusion of C.I. direct orange 34 (MW = 299) and benzoic acid (MW = 122) through degraded semicrystalline polyethylene glycol(PEG)/ poly(L-lactide) (PLLA) block copolymers with various PEG contents and PEG segment lengths at 37°C was studied by UV-visible spectroscopy, differential scanning calorimetry (DSC), wide angle X-ray diffractometer (WAXS), and scanning electron microscopy (SEM). The influences of the

PEG contents, PEG segment lengths, and hydrolytic degradation of PEG/ PLLA copolymers on the solute diffusion coefficient and mode for transport were investigated. It is concluded that the diffusion rate increases with the increase of PEG contents and PEG segment lengths in PEG/PLLA copolymers. It is understandable that the increase of PEG content and PEG segment length both make the degree of crystallinity decrease. The steady state of mass flux could not be reached at the diffusion times up to 1,000 hours, because the copolymers underwent hydrolysis reaction during this period. Furthermore, it was understood that the characteristic time of diffusion as defined by the square of film thickness at an instant of time over the diffusion coefficient of solute through polymer decreased with the increasing diffusion time [66].

The mobility of ethyl acetate (EAc) as a function of crystallinity in poly(vinylidene fluoride) (PVF2) was quantified utilizing methodology based on the recently developed rheo-photoacoustic Fourier transform infrared spectroscopy (RPA FTIR). The diffusion measurements obtained with the photoacoustic cell, combined with the independent sorption and X-ray measurements, indicated that the crystalline phase of the polymer is impermeable to EAc. Spectroscopic RPA FTIR measurements agreed with the independently determined diffusion coefficients and the theoretical predictions relating diffusion to the percent crystalline phase in PVF2. This study indicates that RPA FTIR methodology is highly sensitive to the gas phase analysis and, with the use of proper calibration procedures, can be applied to a quantitative analysis of diffusion in polymers [67].

Crystalline polyhydroxyalkanoates (PHAs), such as poly-D(-)3-hydroxybutyrate (PHB) and its copolymers with poly-3-hydroxyvalerate (P(HB-HV)), were produced by a wide variety of bacteria and had uses in controlled drug delivery systems. The crystallization kinetics and morphology of P(HB-HV) polyesters both with and without the incorporation of a model drug, Methyl Red, were investigated, as they were thought to influence drug-release characteristics. The influence of copolymer composition and incorporation of Methyl Red on radial growth rates (G) of PHB and P(HB-HV) copolymer spherulites was investigated using polarized light video-microscopy. Growth curves could be obtained over a wide range of undercoolings for all the copolymers studied. At a given crystallization temperature, G decreased with increasing HV content and with increasing drug concentration in polymer spherulites. Spherulite morphology appeared to be a complex function of polymer molecular weight, copolymer composition, drug loading, and crystallization temperature T_c. Release of Methyl Red from melt-crystallized matrices of these PHAs was a function of T_c and copolymer composition. Drug release from isothermally crystallized copolymer films occurred progressively more rapidly with increasing

HV content. This could be explained by the progressively poorer drug entrapment within copolymer matrices with increasing HV content. In PHB, Methyl Red was thought to be largely entrapped within spherulites (interlamellar regions), but for copolymers with increasing HV content, progressively greater amounts of drug were excluded at the spherulite surface and at interspherulitic boundaries [68, 69].

Five molecular weight grades of poly(DL-lactic acid) were characterized using gel permeation chromatography, differential scanning calorimetry, and viscometry to determine the effect of molecular weight on the glass transition temperature and the intrinsic viscosity. In addition, dynamic mechanical thermal analysis was used to assess the dynamic storage modulus and the damping factor of the polymer samples by detecting motional and structural transitions over a wide temperature range. Significant relationships were found between the molecular weight and these polymer properties. The five grades of poly(DL-lactic acid) were also incorporated as binders into matrix tablet formulations containing the model drug theophylline and microcrystalline cellulose. Dissolution studies showed significant correlation between the properties of the polymer and the matrix release profiles of the tablets. The release of theophylline slowed down progressively as the polymer molecular weight increased. The differences in release became less significant and reached a limiting asymptotic value as the molecular weight increased to 138,000. Further, tablet index testing was utilized to determine the properties of the polymer granulation. Although there was no correlation with the molecular weight of PLA, brittle fracture index testing indicated very low brittleness for all granulation tested. However, bonding index determinations correlated very well with both the physical-mechanical properties of the polymer and drug release profiles [70].

The evaporation of 2-hydroxy-4-methoxybenzophenone (HMB) from polypropylene was studied by an isothermal method at different concentrations of HMB ranging from 0.3 to 1.2% wt. The evaporation of the additive was correlated according to first-order kinetics, and rate constants of evaporation were calculated. A linear dependence of the rate constants of volatility on the initial concentration of the additive in the polymer was observed. This result suggests that in the given system, volatility is governed by evaporation from the polymer surface. Volatility of HMB from polypropylene depends on the degree of crystallinity of polypropylene. Samples with different degrees of crystallinity were prepared by quenching of the polymer blends and subsequent annealing at 423K. Isothermal measurements at 383 and 423K yielded higher rate constants for annealed samples than for quenched samples. The rate constant of evaporation at 383K was 20% higher for the annealed sample, whereas at 423K it was

27.5% higher. These results are interpreted on the basis of the concentration changes occurring during annealing [71].

An in situ FTIR-ATR method was used to monitor the sorption processes of water and pH 1.3-sulfurous acid in two latex paints and the base polymer common to both. The sorption kinetics could not be described by a simple Fickian model. The spectra also showed evidence of polymer swelling, which was confirmed in separate swelling measurements. Anomalous behavior was noted for the latex paint containing $CaCO_3$ when exposed to sulfurous acid. The amount of water sorbed by this sample went through a maximum, then decreased to a constant level. This was accompanied by similar variations in the degree of swelling of the sample. These changes were explained by the rapid loss of $CaCO_3$ from this particular paint upon exposure to acidic solutions, followed by structural rearrangement to fill in the voids left by the $CaCO_3$ particles [72].

Sorption and diffusion of dichloromethane vapor were measured in amorphous syndiotactic polystyrene films, obtained with different cooling conditions and after controlled aging times at different temperatures. The diffusional behavior, at the temperature of 25°C, was characterized by three stages, depending on penetrant activity. In the first stage, at low activity, the diffusion coefficient was independent of vapor concentration; the second stage was characterized by concentration-dependent diffusion, whereas in the third stage, at high activity, the strong interaction solvent-polymer increased the mobility, allowing the polymer crystallization. The different cooling conditions neither have an effect on the diffusional behavior nor on the sorption curve. The aging, both at room temperature and at 70°C, did not change the diffusion parameters, but led to the appearance of more and more anomalous sorption behavior. The sorption curve, as a function of vapor activity, did not show any difference for the fresh and the aged-at-room temperature samples, whereas the samples aged at 70°C presented a lower sorption at low activity. The presence of ordered domains, impermeable to the penetrant at low activity, was suggested on the basis of sorption results. The solvent-induced crystallization was investigated for all the samples. Crystallization was induced at an activity of 0.45 for the fresh and the aged-at-room temperature samples; at variance, the samples aged at 70°C crystallized at a slightly higher activity, reaching, nevertheless, a higher level of crystallinity [73].

We were modeling the kinetics of water sorption in polymers of structural heterogeneity using Eqs. (2.15)–(2.17) and (2.19) with constants K_{w1}, K_{w2}, and K_{w3} corresponding to the system of poly(hydroxybutyrate) (PHOxB)-water-KCl (NaCl). The computed isotherms of water sorption are shown in Fig. 2.10, and the kinetics of water uptake and drug release are illustrated in Fig. 2.11 If we increase the initial load of salt both

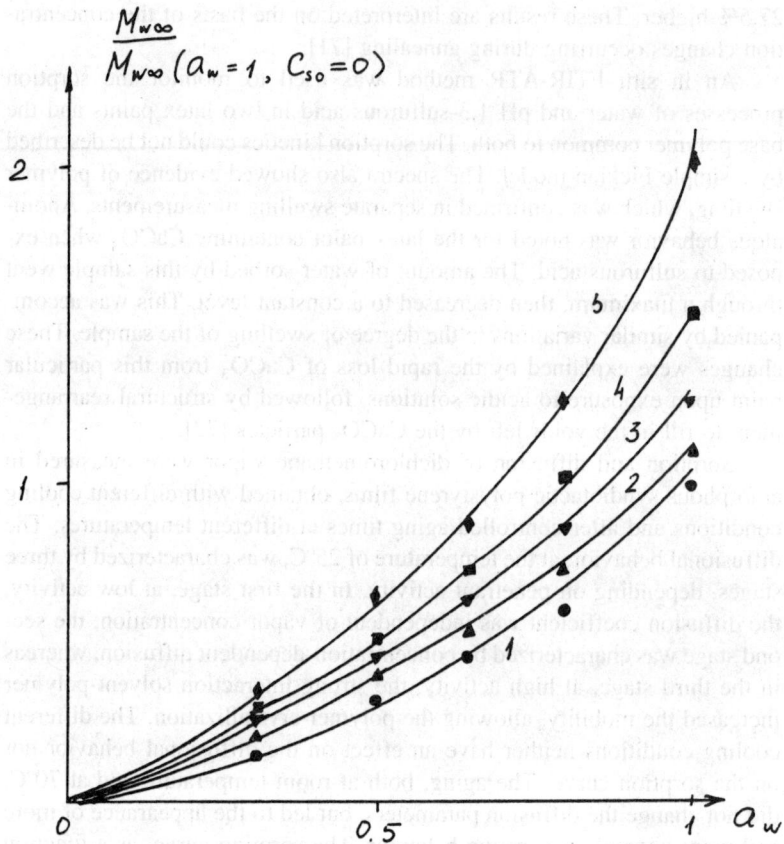

Figure 2.10. Computed and experimental (points) isotherms of water sorption in PHOxB containing KCl.
C_s = 0 (1), 0.1 (2), 0.3 (3), 0.5 (4), 0.8 (5) mol/dm³.

processes accelerate, and starting from the certain value of C_s the kinetic curve of water sorption passes maximum and anomalous character of release is observed. The equilibrium water concentration was found equal

$$C_w = a_v\, K_{w1}\, exp\, (K_{w2}a_w + K_{w3}C_{sim})$$

where C_{sim} is the concentration of the immobile part of the solute. This is also in agreement with experimental results which were obtained for the system of VP/MM copolymer-water-dioxidine.

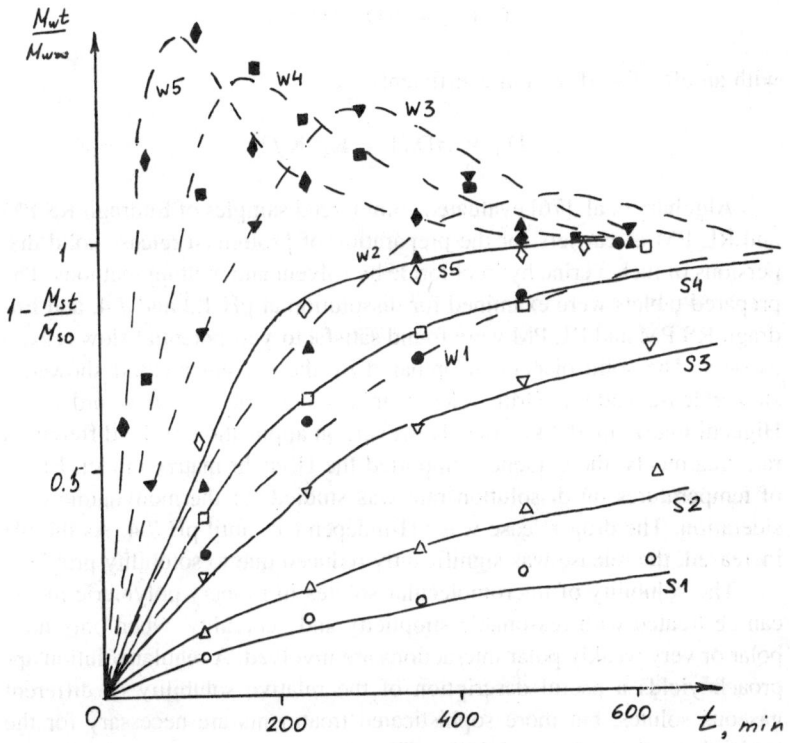

Figure 2.11. Computed and experimental (points) kinetic curves of water sorption and release of potassium chloride from PHOxB films. $C_s = 0$ (1), 0.1 (2), 0.3 (3), 0.5 (4), 0.8 (5) mol/dm^3.

Petropoulos did his computations using Eq. 2.18 instead of 2.19, and got the same qualitative results. He mentioned the intensification of the tendency towards S-shape salt elution curve with the increase of its load. He refered the results of computations to the results reported Lee in [74].

2.2.2. *Effect of the solute nature*

The simple equation for a description of the release of solute of the limited solubility was suggested by Higuchi [75]. This equation is well known as Higuchi kinetics. Using the nomenclature of the current chapter, we can write the Higuchi dependence of the concentration of the released compound on time as

$$C_s/C_{s0} = 2(D_{sF}t/\pi l^2)^{1/2}$$

with an effective diffusion coefficient

$$D_{sF} = \pi D_s (1 - K_s) K_s/2.$$

Algohary et al. [76] evaluated commercial samples of Eudragit RS PM and RL PM as carriers for the preparation of prolonged release solid dispersions of mebeverine hydrochloride by solvent and melting methods. The prepared tablets were examined for dissolution at pH 1.2 and 7.4, and Eudragit RS PM and RL PM were found satisfactory as potential slow release carriers. The solid dispersion prepared by the solvent method showed a slow release pattern. Drug release appeared to fit both first order and Higuchi matrix model kinetics. However, on application of the differential rate treatments, the evidence supported the Higuchi matrix model. Effect of temperatures on dissolution rate was studied for thermodynamic consideration. The drug release was pH-independent until pH 7.4. As the pH increased, the release was significantly reduced due to solubility problem.

The solubility of micromolecular solutes in rubbery polymeric media can be treated with reasonable simplicity and generality, when only nonpolar or very weakly polar interactions are involved. A regular solution approach yields a useful description of the relative solubility of different gaseous solutes, but more sophisticated treatments are necessary for the evaluation of absolute solubility. The treatment of excess solubility in glassy polymers seems to be the topic which currently attracts most interest. The dual mode sorption model, which has long been used for this purpose, as well as alternative approaches recently proposed, were discussed critically in [77]. The effect of pH on absorption of drug and possible reasons for limited absorption are discussed in [78].

The ability to control the fluxes and the relative composition of multiple species permeating across a membrane would be invaluable to the design of controlled release systems. The counter-diffusion of ethanol and water across poly (ethylene-co-vinyl acetate) [P(E-co-VAc)] membranes was investigated as a case study. Ethanol and water fluxes across these membranes were measured from the changes of the refractive indices in the donor and receiver solutions. Ethanol and water solubility in the membranes were determined by a sorption and desorption method. Ethanol absorption followed a Flory-Huggins isotherm, whereas water absorption exhibited a maximum due to opposing factors between the plasticization of the membrane by ethanol and the driving force of water. Ethanol and water diffusivity in the membranes were calculated from the flux and solubility data. In a 37% vinyl acetate P(E-co-VAc) membrane, ethanol permeability was found to increase exponentially with ethanol activity in the membrane; whereas water permeability decreased with water activity in the membrane. Transport models were developed to estimate ethanol and water fluxes across a 37% vinyl acetate P(E-co-VAc) membrane, given any set of compositions in the donor and receiver solutions. The counterdiffusing fluxes of ethanol and water across this membrane were found to be approximately linearly dependent. Additionally, the model suggested that a 37% P(E-co-VAc) membrane slightly favors water over ethanol, in separating ethanol and water by pervaporation [79].

The release of cinnamyl alcohol from spherical polydimethylsiloxane beads of about 1 mm diameter exhibits a considerable zero-order region in pentane. The corresponding release in water is, however, completely Fickian. The observed penetrating front movement and transient dimensional changes in pentane are Fickian-like and, therefore, do not provide sufficient clues to the underlying causes of this anomaly. Deborah number analyses in conjunction with the time-dependent diffusion coefficient approach indicate that both the polymer relaxation in response to solvent swelling and the solute diffusion occur on a comparable time scale, which would suggest a non-Fickian diffusion mechanism. However, the spherical geometry would have rendered the fractional release less linear even under conditions which would normally lead to zero-order release in sheet samples. One possible explanation attributes the apparent zero-order release region to the unusually large swelling of the polymer in pentane (approximately 90% by volume), which causes time-dependent increases in both the surface area as well as diffusion coefficient [80].

Several mathematical models for facilitating the interpretation of the release data of lipophilic drugs from highly lipophilic matrices were proposed. Aminopyrine (AMP), ethinylestradiol (EE), and dipropylphthalate (DPP) were used as model drugs, and acrylic type pressure sensitive adhesive (PSA) was used as a model lipophilic matrix. Release experiments were performed using a diffusion cell at 37°C. AMP was completely released from the PSA, whereas only a fraction of EE and DPP loaded was released. There are usually several resistances involved in the diffusion of a drug through the matrix, such as interfacial transfer resistance (R) and transport resistance across the stagnant layer (H) during the drug-release process; several equations containing above resistance were therefore developed in the Laplace dimension and used to interpret the obtained release data for the three drugs. Analysis of the results suggested that the stagnant layer could be neglected when using a special type of diffusion cell and that the release data of the drugs from the PSA matrix could be interpreted by introducing the interfacial migration resistance and partition coefficient between the matrix and receiver solution into a Fick's diffusion equation [81].

In vitro release kinetics of isoniazid (INH) from a biodegradable polymeric matrix was investigated for systems prepared under varying fabrication conditions. Matrices were prepared either by evaporation of co-solutions followed by extrusion under controlled temperature and pressure, or by dry mixing INH with PLGA which had been preground and sieved to isolate specific particle size ranges prior to extrusion. The effect on release kinetics of loading, polymer particle size, extrusion pressure, and rod diameter was explored. Fractional release as a function of time was shown to fit the Roseman-Higuchi model in that plots of $(1 - F)\ln(1 - F) + F$ are linear with time where F is the fraction of drug released at time t. The effect of varying fabrication parameter values on the slope (identified as the combined rate constant for drug release, K) is as follows: K increases with decreasing polymer particle size, lower extrusion pressure, and lower loading. The effect of increasing the rod diameter results in a small increase in K. Other parameter values being equal, solvent cast systems released more rapidly than those prepared without the use of solvent [82].

Therefore, one of the main problems that occurs is whether the imbibed water (when $C_w = C_{wo}$) can dissolve the whole of the initial salt load of the polymer film or not. Petropoulos [47] did corresponding computations for $k_s = \exp(-k_{ss}/C_w)$. He assumed that k_s is progressively lower as C_w is reduced. For computations more specifically representative of the cellulose acetate-NaCl-water system, he put $k_s = 0.2$ and varied K_s within the realistic frame 0.13–0.013. The results illustrated in Fig. 2.12 show that an S-shape curve is predicted for higher K_s value (higher solubility or lowest solute load), but this S-shape curve tends to disappear as K_s is lowered, and finally the desorption curve obeys the Higuchi kinetics.

This result was obtained for moderate values of $D_s(C_{wo})/D_{wo}$ (of order of 1) and is in contrast with the regularities of solute desorption from swelling matrix (see section 2.1.2) where at equivalent values of $D_s(C_{wo})$ and D_{wo} the solubility affects only solute desorption. Desorption of low-molecular solute with small values of $D_s(C_{wo})/D_{wo}$ (of order of 0.1) becomes Fickian either for swelling or nonswelling polymers due to the fact that water uptake is completed at an early stage.

The swelling and drug release kinetics in suspension polymerized glassy poly(hydroxyethyl methacrylate) beads have been studied in detail using model drugs of both high and low water solubility: oxprenolol HCl (solution. in water is almost equal to 77%) and diclofenac sodium (solution. in water is almost equal to 2.65%). The current results verified the previous findings of Lee and Kim that the drug loading has a definite effect on the drug-release mechanism from hydrogels. The initial swelling front penetration was observed to behave more Fickian as drug loading in-

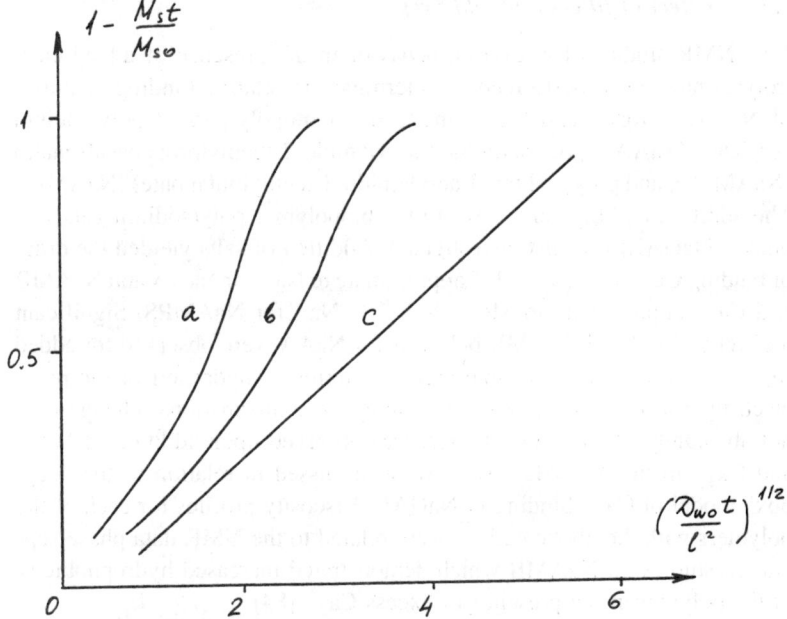

Figure 2.12. Computed salt elution curves.
K_s = 0.13 (a), 0.066 (b), 0.013 (c).

creases. Such a transition can be considered as a change of relative importance of the diffusion process versus the polymer relaxation as a function of drug loading. The swelling front was also observed to accelerate near the core. This was shown to be the natural outcome of the associated moving boundary problem in spherical geometry and not, as proposed by some authors, a super-Case II transport behavior. In all cases, the swelling bead dimension went through a maximum followed by a gradual approach to an equilibrium value during the drug release. In addition, a double-front formation (a swelling front and a drug dissolution front) was observed for diclofenac sodium during the drug release due to the low drug solubility. Lee and Kim also showed that by considering the increase in hydrogel bead dimension due to swelling and the decrease in dimension due to drug release, the swelling maximum in the transient dimensional changes can be qualitatively predicted. The osmotic contributions due to the presence of the drug can then be estimated from the differences between the experimental and calculated transient dimensional changes [83].

2.2.3. Effect of fillers and additives

Na^{23}-NMR studies of relaxation behavior in the presence of added electrolytes have been performed to determine the relative binding affinities
of Na^+ K^+, Mg_2^+, and Ca_2^+ ions to the homopolymers of poly(sodium
acrylate) (NaAA), poly(sodium 2-acrylamido-2-methylpropanesulfonate)
(NaAMPS), and poly(sodium 3-acrylamido-3-methylbutanoate) (NaAMB).
The addition of Mg_2^+ and Ca_2^+ to the biopolymer poly(sodium galactur-
onate) (NaGAL) was also investigated. Addition of salts yielded the order
of binding $Ca_2^+ > Mg_2^+ > K^+$ approximate to Na^+ for NaAA and NaAMB
and Ca_2^+ approximate to $Mg_2^+ > K^+ > Na^+$ for NaAMPS. Significant
differences in the Na^{23}-NMR behavior for NaAA were observed for added
Mg_2^+ and Ca_2^+ and were interpreted in terms of hydration of the polyelectrolyte near phase separation, although a conformational change cannot absolutely be ruled out. Differences observed upon addition of Mg_2^+
and Ca_2^+ in the NaGAL system were discussed in relation to the "egg-
box" model of Ca_2^+ binding to NaGAL. Viscosity profiles for each of the
polymers with the above cations were related to the NMR data phase-sep-
aration studies on NaAMB which demonstrated increased hydrophobicity
of the polymer in the presence of excess Ca_2^+ [84].

Experimental and theoretical permselectivity for a filled elastomeric
membrane had been correlated utilizing Hildebrand solubility parameters
in a solution-diffusion model. Equilibrium partitioning of substances between the feed and membrane phases may be modeled as solutions. These
two phases may be thermodynamically characterized by means of the Hildebrand solubility parameters. In addition, the effect on diffusive transport
of the partitioning of a substance between a filler and polymer phase in a
membrane may be correlated to solubility parameters. A variety of substances were examined using a silicone elastomer membrane, composed of
69% wt poly(dimethylsiloxane) and 31% wt fumed silica. The model was
demonstrated to have good predictive value over three or four orders of
magnitude for the permselectivity of substances including permanent gases,
alkanes, chlorinated and brominated hydrocarbons, alcohols, and aromatic
hydrocarbons [85].

Absorption of water in epoxy adhesives containing different types of
fillers was studied after immersion in distilled water and in NaCl solutions
during several periods of time. The amount of water uptake in the adhesives was found to increase with the concentration of water-soluble fillers
incorporated in the adhesive matrix. The water absorption behavior of the
adhesives investigated was found to be non-Fickian. Owing to reverse osmosis, the amount of water absorbed in the adhesives decreases with the
concentration of the bulk NaCl solution [86].

The physical and mechanical behavior of sisal fiber reinforced polyester composites were studied as a function of sisal fiber content/fiber plies. In order to improve water uptake and wettability characteristics of sisal fibers, N-substituted methacrylamide was used as an interfacial agent. Particulate hybrid composites have been prepared from treated sisal/glass fibers, filler, and unsaturated polyester resin. An improvement in physicomechanical properties of these composites was observed as compared to sisal/polyester and unsaturated polyester resin. A comparison has also been made with the properties of particulate hybrid composites to the commercially available medium density fiberboards and high density particle boards. The new composite has good physico-mechanical properties and performance characteristics [87].

Markin and Zaikov [88] have considered the problem of describing quantitatively the process of diffusion in polymers attended by plasticization of their amorphous phase. A consequence of such plasticization is that the volume accessible to the penetrant molecules continuously increases during the course of their sorption. Assuming that the rate of relaxation processes is far greater than that of diffusion, an equation has been generated to describe the sorption in a matrix with varying accessibility. A numerical solution of this equation, using experimental data for a polyamide-water system, demonstrates that sorption in a matrix with increasing accessibility is slower than sorption in a hypothetical matrix with constant accessibility and with the same proportion of the amorphous phase. At the same time, the concentration of the penetrant in the former case proves to be higher at any point in the polymer specimen. The results obtained are important for calculating the rate constants for chemical reactions that proceed in a polymer matrix in the diffusion-kinetic mode [88].

The interaction of water with poly(vinylchloride) and poly(vinylidene chloride-vinylchloride) (p(VdC-VC)) plasticized with different amounts of a monomeric plasticizer (dioctylphthalate) was studied using inverse gas chromatography [89]. The work was focused on the effect of temperature and plasticizer content on the water sorption behavior of these plastic packaging materials. Values for thermodynamic parameters such as Gibb's free energy (ΔG_s), enthalpy (ΔH_s), entropy (ΔS_s), and activity coefficient (γ), corresponding to sorption of water by the polymers have been calculated using chromatographic retention data. It was found that the sorption of water vapors increases with increasing the amount of plasticizer and decreases with increasing temperature. The Van Deemter equation was found to be applicable to these systems and was used to determine diffusion coefficients and activation energies for diffusion. Diffusion coefficient values increase with increasing amounts of plasticizer and they also increase with increasing temperature; this latter increase is accompanied by a decrease in the activation energy for diffusion [89].

Six potential plasticizers for an ethylcellulose pseudolatex coating system (Aquacoat) were evaluated at three levels (25, 30, and 35%) to study the influence of these additives on the release of a model compound, propanol hydrochloride, from pellets in two different media, dilute HCl and phosphate buffer, pH 7.4. For the majority of the plasticizers, the release rate decreased when larger amounts of plasticizer were incorporated into the coating. However, for the plasticizers dubutyl sebacate and dibutyl adipate, no further reduction in the release rate was observed following dissolution testing in dilute HCl when the level of plasticizer was increased from 30 to 35%. This suggests that the saturation capacity of these plasticizers in the film coating had been exceeded. The media pH was found to influence the dissolution characteristics of the coatings. Faster release rates as well as earlier curve inflection points and T-50% values were observed for plasticizers evaluated in phosphate buffer. All plasticizers used were independent of pH. Correlation of the dissolution results with properties of free films indicated that slower release (more complete film formation) was associated with softer and weaker films with greater elongation [90].

The penetration of water into hydroxypropylmethylcellulose (HPMC)/propranolol matrices containing 25% HPMC K15 and varying quantities of an additive was studied. The kinetics of liquid uptake into these matrices was fitted with a second-order polynomial equation incorporating both Fickian Case I diffusion and Case II relaxational models. Lactose improved the Case I uptake into HPMC/propranolol matrices. However, Case I uptake was reduced in matrices with dicalcium phosphate compared with matrices with lactose. Starch had a minimal effect on the overall water uptake profile. It improved Case I transport marginally but did not affect Case II transport. Sodium carboxymethylcellulose (NaCMC) altered the water penetration profile of HPMC/propranolol matrices two-fold. The presence of 50% NaCMC in the matrices encouraged higher water uptake while matrices with 25% NaCMC increased the water uptake only during the initial stages [91].

The influence of compression force and type of fillers on the release pattern of diclofenac sodium from wet-granulated tablet matrices containing Encompress and lactose as fillers with Eudragit RSPM as a matrix additive was determined. The release patterns of diclofenac sodium from these matrices in USP phosphate buffer pH 7.2 at 37°C were found to be independent of compression forces whereas the ratios of insoluble (Emcompress) to soluble filler (lactose) appeared to greatly influence the drug release rate as evidenced in the increased release rate with decreasing

Emcompress content. The kinetics of drug release was determined and found to precisely obey the Higuchi's planar matrix model. A theoretical approach to predict the drug-release rate from a designed system was presented and found to be satisfactorily predictive [92].

A controlled-release matrix filler was prepared by spray-drying a heated aqueous HPMC solution suspending microcrystalline cellulose (MCC, PH101). Acetaminophen tablets (used as model drug with content of 50%) were prepared by directly compressing the mixture of drug and spray-dried matrix filler. When HPMC was formulated with more than 10% of the matrix filler, drug release from the tablets was satisfactorily sustained. To obtain a similar sustained-release pattern with unmodified original HPMC, more than 50% of the matrix filler was required in the formulation. Whereas, when the tablet formulation was less than 5% modified HPMC, the drug was rapidly released from the tablets. Uniformly distributed HPMC in the spray-dried filler should lead to such drug-releasing behaviors. The micrometric properties of HPMC in the matrix filler, particle size, size distribution, and the loading amount of HPMC were main factors in determining the drug-release properties of the tablets. The drug-release rate of the tablets was determined by the erosion rate of the gelled HPMC formed on the surface of the tablets. The drug-release kinetics were described as a function of the cube root of the tablet weight. The drug was tabletted directly with the modified matrix filler by a rotary tabletting machine [93].

Efentakis et al. [94] studied hydrophobic matrices which were prepared using Eudragit RL 100. Flurbiprofen was used as a model drug, with sorbitol as a diluent. The effect of adding each of five surfactants (sodium lauryl sulphate, sodium taurocholate, cetylpyridinium chloride, cocamidopropyl betaine (CDB), and cetrimide) individually to the matrix was investigated. To investigate the mechanism by which the rate of drug release was increased following the incorporation of surfactants, experiments were undertaken to assess the wettability of the different formulations, and to measure drug release in the presence of submicellar and micellar concentrations of the surfactants. Three mechanisms were proposed by which drug release could be increased following the addition of surfactants: improved wetting, solubilization, and the dissolution of the soluble surfactants to form pores in the matrix. When the surfactant was added to the dissolution fluid, only one surfactant (CDB) did not result in an increase in drug release; for the other surfactants a minor increase in drug release was observed. Therefore, in most cases, wetting plays a small role in aiding dissolution. There was no significant change in release rate when the experiment was performed in the presence of either sub-micellar or micellar concentrations of the surfactants, thus solubilization of the drug does not seem to be implicated in the drug-release mechanism. The most significant increase in drug-release rate was caused by incorporating the most soluble surfactants (sodium taurcholate and cetrimide) within the matrix. As the increase was significantly greater than could be explained by wetting alone, it must be concluded that for these matrix systems the major mechanism by which surfactants increase the dissolution rate is by the formation of pores to aid the access of the dissolution fluid and egress of the dissolved drug. It is also possible that the presence of the relatively concentrated surfactant solution in the wetted tablet would reduce interparticle adhesion and thereby speed drug-release rate as a result of an increased disintegration [94].

The permeability of urea, sucrose, L-alanine, and their mixtures through a cellulose (Cuprophane) membrane and a copolyether-urethane membrane based on polyoxyethylene glycol was measured in [95]. It was found that solute diffusion through the copolymer membrane was affected by the presence of a second solute. This effect appears to be related to the water-structure breaking (with urea) or water-structure making (with sucrose) nature of the co-solute with the water/polyoxyethylene phase of the copolymer membrane. This effect was not seen for the cellulose membrane.

Electrolytes are often added to a gel-swelling medium under the assumption that the important conditions which characterize swelling rates are the solution pH and ionic strength, with little emphasis on the nature of the electrolyte. Previous research by Siegel et al. [96] indicated that the presence of the un-ionized acidic form of an electrolyte buffer is a primary rate determinant for swelling of a polybase gel. A systematic swelling study on two separate gels, 2-hydroxyethyl methacrylate copolymerized with methacrylic acid (HEMA/MAA) and N,N-dimethylaminoethyl methacrylate (HEMA/DMA), was performed to investigate the influence of the concentration of the un-ionized buffer by three principal factors: (i) total buffer concentration, (ii) solution pH, and (iii) buffer pK. Swelling and deswelling kinetics were obtained. In the presence of an electrolyte buffer, a dramatic swelling rate increase was observed for the HEMA gels, with substantial gains in rate obtained as total buffer concentration rises. Results obtained in [97] also emphasize that to enhance swelling kinetics, the pH must be such that the buffer is essentially un-ionized.

Nguyen et al. presented a method for studying anomalies in the diffusion of gases and vapors in dense polymer films in the transient regime. It was applied to the diffusion of CO_2 and water vapor in plasticized-poly (vinylchloride) (PVC)-potato-starch blends. The apparent diffusion coefficient of CO_2 in plasticized PVC remains constant in the transient regime, while that of water vapor first increases, passes through a maximum, and then finally decreases to a constant value. The anomalous process observed for water vapor permeation in the transient regime could be explained by the interaction between the diisooctylphthalate (DIOP) plasticizer and the deformation of the polymer network due to stresses generated by the starch particles swelling. For water vapor in the stationary regime, the diffusion coefficient decreased and sorption coefficient increases when the starch content increases. There was a strong increase in both sorption and diffusion coefficients when pure PVC was replaced by DIOP-plasticized PVC [98].

The effect of oxprenolol HCl loading on the kinetics of polymer swelling and drug release from suspension-polymerized poly(methyl methacrylate-co-methacrylic acid) beads was studied in detail by Kim and Lee [108]. Within the range of variables studied, the polymer swelling rate increases with buffer pH and concentration. And an ionization-controlled swelling mechanism (analogous to the relaxation-controlled mechanism) was more rate-limiting at higher buffer concentrations. At oxprenolol HCL loading levels below 17.8%, the drug release and associated dimensional changes (in pH 7.4) exhibited an extended quasi-linear region despite the inherent limitation of spherical geometry. At higher loading levels, the drug release becomes faster and first-order in nature. This is apparently a result of the transition from a dissolved to a dispersed system above a percolation threshold (15–18% loading in the present study). As a result of competition from processes such as the reduction of bead dimension due to drug release and the dimensional increase due to polymer swelling and osmotic contributions from the drug, the transient bead diameter increased monotonically during drug release at loading levels up to 25.6%, whereas upon further increasing the drug loading, the bead diameter went through a maximum during the early drug release, which eventually increases again as a result of the slow but continuous increase in polymer swelling due to further ionization. In all cases, both the drug release and the dimensional changes approach completion as the penetrating ionization fronts met at the center, indicating a true swelling-controlled behavior [99].

The effect of the third component depends on its own solubility in water. Obviously, the corresponding equation should be added in the system Eqs. (2.15)–(2.17) or (2.3)–(2.11)

$$\delta C_f/\delta t = (\delta/\delta x)(D_f S_f \ \delta a_s/\delta x)$$

and the dependencies of solubility of water, and, in general, elastic modules and frequency of relaxation should contain the third contributing factor.

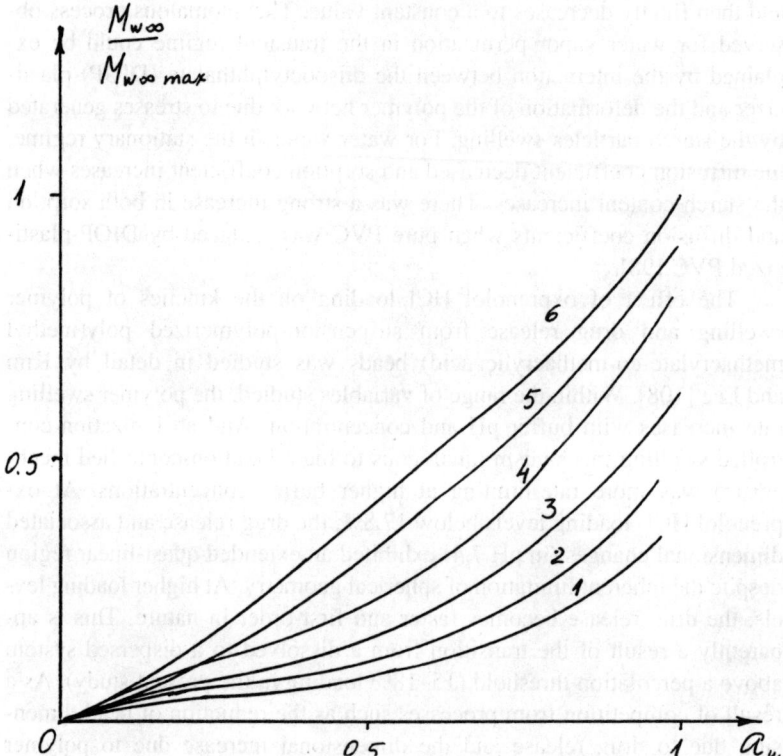

Figure 2.13. Computed water isotherms in polymer containing highly and lower soluble solutes.
C_F/C_s = 0.1 (1), 0.2 (2), 0.3 (3), 0.5 (4), 0.8 (5), 1 (6).

The simplest cases are when the third component just make the aforesaid dependencies stronger or weaker. Then the main regularities described above remain unchanged. There is, however, a remarkable possibility to alter the hydrophilic character of the polymer in principle.

The effect of plasticizer was modeled by the addition to the system of an equation and the replacement of Eq. (2.19) by

$$S_w = C_w/a_w = K_{w1} \exp (K_{w2}a_w + K_{w3}C_s + K_{w4}C_f)$$

and the effect of concentration of this filler has been investigated. The results are shown in Figs. 2.13 and 2.14. Up to a certain value the increase of C_f accelerates water sorption and the release of solute, and their kinet-

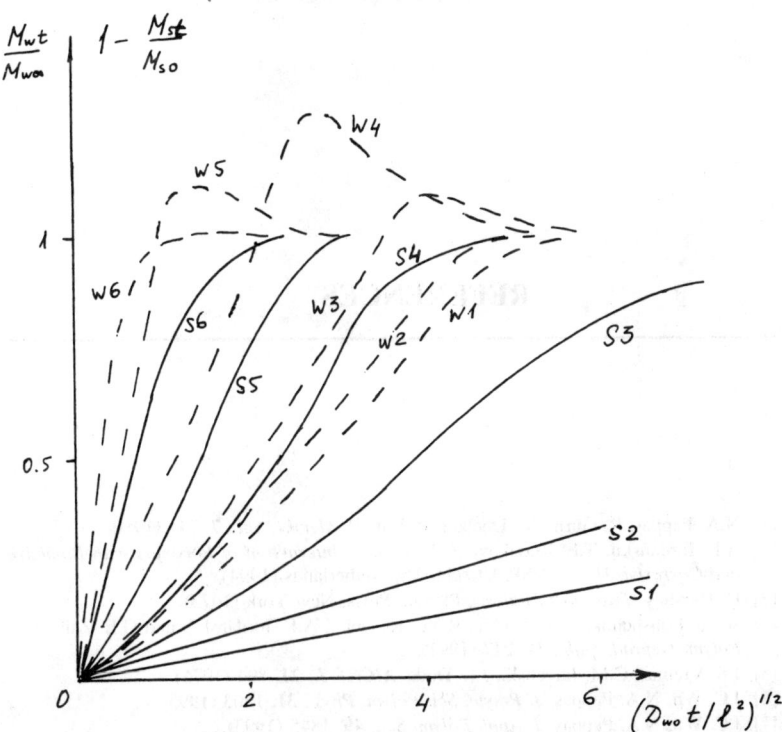

Figure 2.14. Computed kinetics of water sorption in and solute release from polymer filled with the third component.
C_F/C_s = 0.1 (1), 0.2 (2), 0.3 (3), 0.5 (4), 0.8 (5), 1 (6).

ics and shape of the water sorption isotherm remain unchanged. Then, however, remarkable changes come in. This effect may be explained by the increase of the solubility of solute. Accordingly, the contribution of solute in the solubility of water becomes negligible, and this changes the shape of water isotherm. Further acceleration of solute release also accompanies the change in its kinetics.

REFERENCES

[1] N.A. Peppas, R. Gurny, E. Doelker, P. Buri. *J. Membr. Sci.,* **7**, 241 (1980).
[2] A.L. Iordanskii, T.E. Rudakova, G.E. Zaikov. *Interaction of Polymers with Bioactive and Corrosive Media.* VSP, Utrecht, The Netherlands (1994).
[3] D. Hershey. *Transport Analysis.* Plenum Press, New York (1973).
[4] A.Ya. Polishchuk, L.A. Zimina, R.Yu. Kosenko, A.L. Iordanskii, and G.E. Zaikov. *J. Polym. Degrad. Stab.,* **31**, 247 (1991).
[5] J.S. Vrentas, C.M. Jarzebski, J.L. Duda. *AICHE J.,* **21**, 894 (1975).
[6] J.C. Wu, N.A. Peppas. *J. Polym. Sci. Polym. Phys.,* **31**, 1503 (1993).
[7] J.C. Wu, N.A. Peppas. *J. Appl. Polym. Sci.,* **49**, 1845 (1993).
[8] A.T. Pham, P.I. Lee. *Pharm. Res.,* **11**, 1379 (1994).
[9] J.H. Petropoulos, P.R. Roussis. *J. Membr. Sci.,* **3**, 343 (1978).
[10] J.H. Petropoulos. *J. Polym. Sci. Polym. Phys. Ed.,* **22**, 183 (1984).
[11] A.Ya. Polishchuk, G.E. Zaikov, J.H. Petropoulos. *Inter. J. Polymer. Mater.,* **25**, 1 (1994).
[12] A.Ya. Polishchuk, G.E. Zaikov, J.H. Petropoulos. *Inter. J. Polymer. Mater.,* **19**, 1 (1993).
[13] A.Ya. Polishchuk et al. *Polymer Sci. USSR,* **32A**, 2203 (1990).
[14] H. Yasuda, C.E. Lamaze, A. Peterlin. *J. Polym. Sci. A2,* **9**, 1117 (1971).
[15] L.A. Zimina, A.Ya. Polishchuk, N.N. Maduskin, A.L. Iordansky, G.E. Zaikov. *Intern. J. Polymeric Mater.,* **16**, 185 (1992).
[16] J.H. Petropoulos, K.G. Papadokostaki, S.G. Amarantos. *J. Polym. Sci. Polym. Phys.,* **30**, 717 (1992).
[17] M. Sanopoulou, J.H. Petropoulos. *J. Non-Crystalline Solids,* **131**, 827 (1991).
[18] M. Sanopoulou, J.H. Petropoulos. *J. Polym. Sci. Polym. Phys.,* **30**, 983 (1992).
[19] M. Sanopoulou, J.H. Petropoulos. *J. Polym. Sci. Polym. Phys.,* **30**, 971 (1992).
[20] S. Govindjee, J.C. Simo. *J. Mech. Phys. Solids,* **41**, 863 (1993).
[21] F. Doghieri, G.C. Sarti. *Makromol. Chemie - Macromol. Symp.,* **68**, 257 (1993).
[22] C.K. Hayes, D.S. Cohen. *J. Polym. Sci. Polym. Phys.* **30**, 145 (1992).
[23] D.S. Cohen, A.B. White. *SIAM J. Appl. Math.,* **51**, 472 (1991).
[24] E. Sancaktar, D.R. Baechtle. *J. Adhesion,* **42**, 65 (1993).
[25] D. Hariharan, N.A. Peppas. *J. Polym. Sci. Polym. Phys.,* **32**, 1093 (1994).
[26] D. Hariharan, N.A. Peppas. *J. Controlled Release,* **23**, 123 (1993).
[27] D.J. Kim, J.M. Caruthers, N.A. Peppas, *Polymer,* **34**, 3638 (1993).
[28] M.C. Lee, N.A. Peppas. *J. Compos. Mater.,* **27**, 1146 (1993).
[29] K.C. Sung, E.M. Topp. *J. Membr. Sci.,* **92**, 157 (1994).

[30] A.L. Iordansky, A.Ya. Polishchuk, R.Yu. Kosenko et al. *Inter. J. Polymer. Mater.*, **16**, 195 (1992).
[31] R. Gref et al. *J. Appl. Polym. Sci.* **49**, 209 (1993).
[32] M.E. McNeill, N.B. Graham. *J. Biomater. Sci. Polym. Ed.*, **5**, 111 (1993).
[33] B.I. Dasilveira. *Europ. Polym. J.*, **29**, 1095 (1993).
[34] J.H. Jou et al. *J. Appl. Polym. Sci.*, **5**, 857 (1993).
[35] J.G. Vanalsten, J.C. Coburn. *Macromolecules*, **27**, 3746 (1994).
[36] N.A. Peppas, A.R. Khare. *Adv. Drug Deliv. Rev.*, **11**, 1 (1993).
[37] M.E. McNeill, N.B. Graham. *J. Biomater. Sci. Polym. Ed.*, **4**, 305 (1993).
[38] M. Zulfiqar, A. Qudos, S. Zulfiqar. *J. Appl. Polym. Sci.*, **51**, 2001 (1994).
[39] H.A. Davis. *Textile Chemist and Colorist*, **24**, 19 (1992).
[40] R.A. Siegel. *J. Phys. Chem.*, **95**, 2556 (1991).
[41] P. Defilippis et al. *Drug Develop. Industr. Pharm.*, **17**, 2017 (1991).
[42] G. Buckton, M. Efentakis, Z. Hussain. *Europ. J. Pharm. Biopharm.*, **37**, 154 (1991).
[43] M.D. Blanco, J.M. Rego, M.B. Huglin. *Polymer*, **35**, 3487 (1994).
[44] P. Colombo et al. *Intern. J. Pharm.*, **88**, 99 (1992).
[45] N.A. Peppas, J. Klier. *J. Controlled Release*, **16**, 203 (1991).
[46] A.Ya. Polishchuk, G.E. Zaikov, J.H. Petropoulos. *Proc. Diffusion in Polymers -IV, Polymat' 94*, Imperial College, London (1994).
[47] S.G. Amarantos, K.G. Papadokostaki, J.H. Petropoulos. *Characterization of Radioactive Waste Forms*. Final Report for the Commission of the European Communities, Luxembourg (1991).
[48] A.R. Khare, N.A. Peppas. *J. Biomater. Sci. Polym. Ed.*, **4**, 275 (1993).
[49] T. Ostberg, E.M. Lund, C. Graffner. *Intern. J. Pharm.*, **112**, 241 (1994).
[50] M. Ashraf et al. *Pharm. Res.*, **11**, 733 (1994).
[51] E. Foldes, B. Turcsanyi. *J. Appl. Polym. Sci.*, **46**, 507 (1992).
[52] A. Heintz, W. Stephan. *J. Membr. Sci.*, **89**, 143 (1994).
[53] T.B. Meluch, D.R. Lloid. *Polymer*, **34**, 1984 (1993).
[54] X.Y. Su, R. Alkassas, A.L.W. Po. *Europ. J. Pharm. Biopharm.*, **40**, 73 (1994).
[55] G. Zimgone, F. Rubessa. *STP Pharma Sci.*, **4**,122 (1994).
[56] J.H. Guo. *Drug Develop. Industr. Pharm.*, **20**, 2467 (1994).
[57] R.W. Korsmeyer, N.A. Peppas. *Proc. Intern. Symp. on Controlled Release of Bioactive Mater.*, **10**, 141 (1983).
[58] R.W. Korsmeyer, S.R. Lustig, N.A. Peppas. *J. Polym. Sci. Polym. Phys.*, **24**, 395 (1986).
[59] A. Apicella, H.B. Hopfenberg. *J. Appl. Polym. Sci.*, **27**, 1139 (1982).
[60] J.H. Petropoulos, P.P. Roussis. *J. Chem. Phys.*, **47**, 1491 (1967).
[61] A.Ya. Polishchuk, L.A. Zimina, G.E. Zaikov. *Polymer Science* (1995), in press.
[62] W.M. Saltzman et al. *J. Appl. Polym. Sci.*, **48**, 1493 (1993).
[63] Y.P. Ageev, N.N. Matushkina, N.L. Strusovskaya. *J. Membr. Sci.*, **67**, 167 (1992).
[64] A.V. Popoola. *J. Appl. Polym. Sci.*, **49**, 2115 (1993).
[65] R. Gobel et al. *Appl. Spectroscopy*, **48**, 678 (1994).
[66] H.J. Liu, C.T. Hsieh, D.S.G. Hu. *Polymer Bulletin*, **32**, 463 (1994).
[67] B.W. Ludwig, M.W. Urban. *Polymer*, **34**, 3376 (1993).
[68] S. Akhtar, C.W. Pouton, L.J. Notarianni. *Polymer*, **33**, 117 (1992).
[69] S. Akhtar, C.W. Pouton, L.J. Notarianni. *J. Controlled Release*, **17**, 225 (1991).
[70] M.O. Omelczuk, J.W. McGinity. *Pharm. Res.*, **9**, 26 (1992).
[71] J. Luston, V. Pastusakova, F. Vass. *J. Appl. Polym. Sci.*, **48**, 219 (1993).
[72] C.M. Balik, J.R. Xu. *J. Appl. Polym. Sci.*, **52**, 975 (1994).
[73] V. Vittoria, A.R. Filho. *J. Appl. Polym. Sci.*, **49**, 247 (1993).
[74] P.I. Lee. *Polymer*, **24**, 45 (1983).
[75] T. Higuchi. *J. Pharm. Sci.*, **52**, 1145 (1963).
[76] O.M.N. Algohary, S.S.A. Elhady, N.A. Daabis. *Drug Develop. Industr. Pharm.*, **17**, 2055 (1991).

[77] J.H. Petropoulos. *Pure & Appl. Chem.,* **65**, 219 (1993).
[78] B. Hirst. *Proc. Seminar on Advances in Drug Delivery Systems,* Centre in Scotland on
      Nanotechnology, Glasgow (1994).
[79] S.M. Dinh, B. Berner, Y.M. Sun, P.I. Lee. *J. Membr. Sci.,* **69**, 223 (1992).
[80] M.H. Litt et al. *J. Polym. Sci. Polym. Chem.,* **31**, 183 (1993).
[81] T. Kokubo, K. Sugibayashi, Y. Morimoto. *J. Controlled Release,* **20**, 3 (1992).
[82] Y.Y. Hsu et al. *J. Controlled Release,* **31**, 223 (1994).
[83] P.I. Lee, C.J. Kim. *J. Controlled Release,* **16**, 229 (1991).
[84]   J.K. Newman, C.L. McCormick. *Macromolecules,* **27**, 5114 (1994).
[85] M.A. Lapack et al. *J. Membr. Sci.,* **86**, 263 (1994).
[86] R.C.L. Tai, Z. Szklarskasmialowska. *J. Mater. Sci.,* **28**, 6199 (1993).
[87] B. Singh et al. *Research and Industry,* **39**, 38 (1994).
[88] V.S. Markin, G.E. Zaikov. *Polym. Degrad. Stab.,* **40**, 395 (1993).
[89] P.J. Kalaouzis, P.G. Demertzis. *Polymer International,* **32**, 125 (1993).
[90] D.E. Hutchings, A. Sakr. *J. Pharm. Sci.,* **83**, 1386 (1994).
[91] L.S.C. Wan, P.W.S. Heng, L.F. Wong. *STP Pharma Science,* **4**, 213 (1994).
[92] N. Sarusita, P. Mahahpunt. *Drug Develop. Industr. Pharm.,* **20**, 1049 (1994).
[93] Y. Kawashima et al. *Chem. Pharm. Bulletin,* **41**, 2156 (1993).
[94] M. Efentakis et al. *Intern. J. Pharmaceutics,* **70**, 153 (1991).
[95] D.H.T. Lee, D.J. Lyman. *Intern. J. Polymer. Mater.,* **21**, 199 (1993).
[96] R.A. Siegel et al. *J. Controlled Release,* **8**, 179 (1988).
[97] L.Y. Chou et al. *J. Appl. Polym. Sci.,* **45**, 1411 (1992).
[98] X.Q. Nguyen, M. Sipek, Q.T. Nguyen. *Polymer,* **33**, 3698 (1992).
[99] C.J. Kim, P.I. Lee. *Pharm. Res.,* **9**, 1268 (1992).

3 TRANSPORT DEVICES FOR CONTROLLED DELIVERY

3.1. RESERVOIR DEVICES

3.1.1. Diffusion controlled transport

Diffusion controlled reservoir systems include an internal core coated with a membrane (usually thin). A core contains a drug or other chemical and, if required, filaments, osmotic agent, solvent, etc. If a membrane is not degradable, the kinetics and the rate of release are controlled by diffusion and, accordingly, the processes which affect diffusion. The diffusion of solute from the core of different geometry through the polymer membrane is described by the general equation

$$\delta C_s / \delta t = \text{div } J_s \qquad (3.1)$$

where C_s and J_s are the solute concentration in and its flux through the membrane.

$$J_s = -D_s \text{ grad } C_s \qquad (3.2)$$

The most typical boundary conditions for Eq. 3.1 are

$$C_s = k_s c_s; \ (V/A)\delta c_s / \delta t = -J_s \qquad (3.3)$$

at the core/membrane interface (V and A are the volume and the surface area of the reservoir, respectively, and c_s is the concentration of the aqueous solution of the solute in the core), and

$$C_s = C_{s1}, \sim 0 \qquad (3.4)$$

at the membrane/solvent interface.

In assumption of c_s = constant = c_{s0} and constant value of the diffusion coefficient (D_s) we get the expressions which describe the steady-state release from reservoirs of different geometry [1]:

$$M_s = D_s k_s A c_s, t/l \quad \text{(for a plate membrane of the thickness l)} \qquad (3.5)$$

$$M_s = D_s k_s A c_s, t/\ln (R_e/R_i) \quad \text{(for a hollow cylinder)} \qquad (3.6)$$

$$M_s = 4\pi D_s k_s A c_s R_e R_i/(R_e - R_i) \quad \text{(for a hollow sphere)} \qquad (3.7)$$

where M_s is the quantity of the substance to be released. R_e and R_i are the internal and external radii of a cylinder or sphere.

Expressions (3.5)–(3.7) correspond to the constant rate of release which can be obtained for any geometry of reservoir for a case of Fickian diffusion of solute through the membrane. This is the principle difference between this case and Case-II anomalous diffusion in polymeric monoliths.

The rate of release through coatings can be altered by the change in hydrophobic/hydrophilic balance of the membrane. For example, the acceleration of release may be reached by the use of polymer-solute system of the similar polarity [1] as it was obtained in [2, 3].

The process of diffusion controlled release is one of the membrane separation methods which are used as an alternative to conventional separation methods. Various mechanisms have been distinguished to describe the transport in membranes: transport through bulk material (dense membranes), Knudsen diffusion in narrow pores, and viscous flow in wide pores or surface diffusion along pore walls. In practice, the transport can be a result of more than only one of these mechanisms. All of these mechanisms models were derived in [4]. The characteristics of a membrane, e.g., its crystallinity or its charge, can also have major consequences for the rate of diffusion in the membrane, and hence for the flux obtained. Apart from the diffusion transport processes in membranes mentioned above, other important diffusion processes are related to membrane processes, for example, diffusion in the boundary layer near the membrane (concentration polarization phenomena) and diffusion during membrane formation. The degree of concentration polarization is related to the magnitude of the mass transfer coefficient which, in turn, is influenced by the diffusion coefficient. The effect of concentration polarization can be rather different for the various membrane processes. The phase inversion membrane formation mechanism was determined by Vandenberg and Smolders to a large extent by the kinetic aspects during membrane formation, which were diffusion of solvent and of nonsolvent and the kinetics of the phase separation itself [4].

A new anionic composite bead system with a transient membrane-matrix structure, capable of prolonged constant-rate drug release, was developed from suspension-polymerized poly(methyl methacrylate-co-methacrylic acid) (PMMA/MAA). These composite beads had a thin PMMA/MAA surface layer and a core consisting of the sodium salt form of the polymer (PMMA/MANa). The high loading (> 20%) of a model drug (oxprenolol HCl) that was achievable in this system from a loading solution concentration as low as 0.5% suggested the formation of a drug-polymer complex in the form of an ionic salt in the core. The release of oxprenolol from such composite beads showed an initial burst effect followed by an extended constant-rate region before leveling off. Apparently, the surface PMMA/MAA layer functioned as a transient rate-controlling membrane before it was completely ionized. Because the process of ionization was slow, the rate-controlling characteristics of the surface layer and the resulting constant rate of drug release were both sustained for an extended period. The unique feature of this system was not only its high drug loading capability, but also the transient nature of the rate-controlling surface layer, which was completely ionized towards the latter part of the drug release, thus avoiding prolonged tailing of drug release that was normally associated with permanent membrane-matrix systems [5].

The applicability of membrane-reservoir devices based on cold water insoluble, fully hydrolyzed poly(vinyl alcohol) (PVA) membranes for the controlled release of chlorinated isocyanurates into aqueous media was demonstrated in [6]. In addition to providing a prolonged constant rate of release, fully hydrolyzed PVA membranes also exhibited good stability in the presence of saturated solutions of chlorinated isocyanurates. The apparent resistance to degradation by chlorine and the significant retention of wet mechanical strength of these membranes were attributed to the semi-crystalline nature of PVA.

Highly crystalline porous hollow poly (L-lactide) fibers suitable for the delivery of various drugs were obtained using a dry-wet spinning process. The pore structure of the fibers was regulated by changing the spinning systems and spinning conditions [7]. These fibers are suitable for the long-term zero-order delivery of the contraceptive 3-ketodesogestrel and the short-term zero-order delivery of the cytostatic agent, cisplatin. The drugs were released by dissolution of the drug crystals in the fiber core followed by diffusion through the membrane structure. Short-term release of adriamycin could be obtained through an adsorption-desorption mechanism. The pore structures of the fibers have a large influence on the release rates of the drugs investigated. When fibers with dense top layers were used, low release rates of drugs were observed, whereas fibers with well interconnected pore structures over the fiber wall showed very high release rates.

Porous ethylcellulose (EC) film coating technique was used in preparing a capsule-type controlled release dosage form, in which theophylline (TP) was used as a poorly water-soluble model drug. The TP-loaded uncoated beads were spray-coated with an aqueous ethanolic or ethanolic EC solution, and the drug-release characteristics and productivity of each product were examined [8]. When the aqueous ethanol was used as the solvent of the coating solution, a large number of micropores were formed in the coating, and the porosity of coating and drug-release rate could be controlled by altering the ethanolic concentration in the coating solution. In addition, few agglomerates were produced in the coating process, even though there was no anti-agglomeration agent in the coating solution. The drug-release rate from the coated beads could be changed by film porosity as well as film thickness. Superposition analysis revealed that the EC-coated beads with different film porosity or different coating levels had the same drug-release mechanism. It was further found that the drug-release behavior of the porous EC film-coated beads was not affected by any simulated physiological conditions such as pH, surface tension, ionic strength, or paddle rotation speed, indicating that in vivo drug release should not be affected by such physiological conditions in the gastrointestinal tract [8].

Hydrogels were synthesized as the drug reservoir matrix for peptide-based pharmaceuticals, and the iontophoretic release and transdermal delivery of three model peptides, insulin, calcitonin, and opressin, from these hydrogel-based iontotherapeutic devices were investigated by Banga and Chien [9]. The swelling behavior of polyacrylamide-type hydrogel as a function of its monomer and cross-linker concentration was studied, and a hydrogel with minimal swelling was synthesized. The release of peptides from the hydrogel matrix was found to follow a M_s vs $t^{1/2}$ relationship under passive diffusion conditions, which shifted to a M_s vs t relationship under iontophoresis-facilitated transport. The release flux (d M_s/dt) of peptides was observed to decline when the electric current was turned off and was resumed when the current was turned on, thus allowing for modulation of drug release by varying the application parameters of iontophoresis-facilitated transport. The permeability coefficients for these peptides across the hairless rat skin were evaluated using the hydrogel formulations prepared from polyacrylamide, p-HEMA, and carbopol. A rank order of vasopressin > calcitonin > insulin was obtained in accordance with the order of molecular size [9].

NMR microscopy may be used to monitor the formation of the gel layer in hydrating hydrophilic polymer tablets. Bowtell et al. [10] found that the rate and extent of the swelling of the outer gel layer critically influences the kinetics of drug release. Tablets were hydrated in distilled water at 37°C and then imaged at discrete time intervals using a 500 MHz microscope. The growth of the gel layer was clearly observed in time sequences of radial and axial sections. Axial images showed some interesting dimensional changes, with the gel at the flat surface of the tablet developing a concave shape. This was probably a response to the occurrence of uniaxial stress relaxation as hydration proceeds the occurrence of uniaxial stress relaxation as hydration proceeds. Diffusion- and T_2-weighted images provided evidence that the water in the gel layer is more strongly bound close to the dry core of the tablet than at the more fully hydrated outer surface. In images of tablets containing diclofenac, disruption of the gel layer was shown to occur primarily from the flat surfaces of the tablet, while the distribution of particles could be seen in tablets doped with insoluble calcium phosphate [10].

A novel controlled drug-release system, Time-Controlled Explosion System (TES), had been developed by Ueda et al. TES had a four-layered spherical structure, which consists of core, drug, swelling agent, and water insoluble polymer membrane. TES was characterized by a rapid drug release with a precisely programmed lag time; i.e., expansion of the swelling agent by water penetrating through the outer membrane, destruction of the membrane by stress due to swelling force and, subsequent rapid drug release. For establishing the concept and development strategy, TES was designed using metoprolol and polystyrene balls (size: 3.2 mm in diameter) as a model drug and core particles. Among the polymers screened, low-substituted hydroxypropylcellulose and ethylcellulose (EC) were selected for a swelling agent and an outer water insoluble membrane, respectively. The release profiles of metoprolol from the system were not affected by the pH of the dissolution media. Lag time was controlled by the thickness of the outer EC membrane; thus, a combination of TES particles possessing different lag times could offer any desired release profile of the model compound, metoprolol [11].

3.1.2. Osmosis-activated transport

Osmosis is a process in which the solvent is transported through the membrane as a result of a difference in trance-membrane concentration

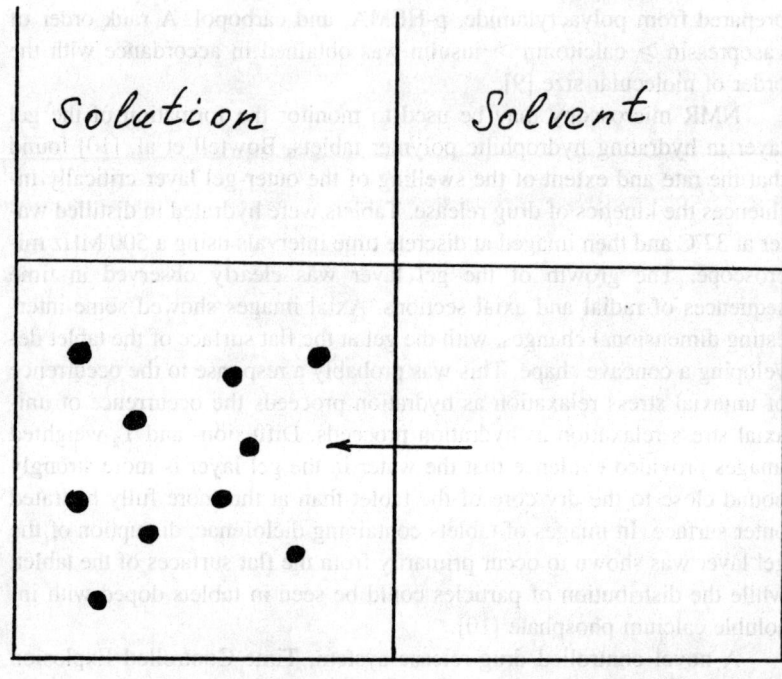

Figure 3.1. Osmosis: solvent flux (shown by arrow) as a result of difference in trans-membrane concentration.

(Fig. 3.1) [12]. If the system is not subject to any external influence, the hydrostatic pressure difference is established, at which point net material point has reached zero. This condition is known as osmotic equilibrium, and the corresponding pressure difference is referred to as osmotic pressure [12]. Osmotic equilibrium is a hydrodynamic equilibrium: solvent still passes through the membrane, but fluxes are statistically the same in both directions. The definition equation for osmotic pressure (π_i) in the case which pure solvent is present on the one side of the membrane is

$$\pi_i = -(RT/V_i) \ln a_i$$

where i is the solvent (e.g., water).

For the limiting case of an ideal dilute solution, Van't Hoff (1885) proposed the equation

$$\pi_w = c_s RT$$

but then established that it described well the osmotic behavior of non-dissociated solutions (organic compounds), while the calculated osmotic pressures were too low for electrolytes (strong acids, strong bases, and most salt solutions). Consequently, he introduced the empirical coefficient β and he wrote

$$\pi_i = -\beta(RT/V_i) \ln a_i.$$

Arrhenius (1887) provided a theoretical explanation of this equation by showing that anomalous osmotic effects of the electrolytes are caused by an increase in the number of osmotically effective particles by dissociation.

The osmosis is applied in most reservoir devices for controlled delivery to establish and govern the kinetics of water transport from external phase to the core. This effect becomes especially substantial in devices with a hole in the coating, which are known as Oral Osmotic Systems (OROS) or Osmotic Pumps. In contrast with diffusion controlled reservoir systems, here we get a case when release of a substance from the core obeys the mechanism of convection and is controlled by water diffusion.

We obtained experimental justifications for a diffusion-convective model of solvent transport and solute release in osmotic pumps when we studied an OROS-system for controlled release of sodium oxacillin, initially granulated with lactose and polyvinylpirrolydone and film-coated with cellulose diacetate membrane [13]. Then the model was developed and applied for analysis of other osmotic polymer systems [14, 15].

Generally, the diffusion-kinetic process in the aforesaid systems is described by the following equations

$$\delta C_w/\delta t = r^{-2} \, \delta/\delta r [D_w \, r^2 \, \delta C_w/\delta r] \qquad R < r < R + 1 \qquad (3.8)$$

$$C_{w0} = C_0 \, [1 - \exp(-k_w t)] \qquad R = r \qquad (3.9)$$

$$C_{w1} = {}_0\!\int^t C_1(\pi_w) \exp[-k_s(t - \tau)]d\tau \qquad R = R + 1 \qquad (3.10)$$

$$c_s = \exp[(\sigma/\rho V_0)_0\!\int^t j_w \, d\tau - {}_0\!\int^t v_d(\tau) \, d\tau] \qquad (3.11)$$

$$j_s = j_w \, C_s/\rho \qquad (3.12)$$

$$v_s = j_s \, \sigma \qquad (3.13)$$

where C_w is the concentration of solvent in polymer, C_{w0} and C_{w1} are its boundary concentrations in the interfaces with liquid solvent and saturated salt solution, respectively. D_w is the diffusion coefficient of solvent (water). C_0 and C_1 are C_{w0} and C_{w1} values in equilibrium (C_0 and C_1 correspond to the solvent activity in a liquid state or in saturated salt solution), c_s is the concentration of the saturated salt solution, k_w and k_s are the rate constants of the process of boundary relaxation, j_w and j_s are the water and solute fluxes, respectively, v_d is the velocity of core dissolution, ρ is the solvent density, σ is the cross-area of the hole, v_s is the velocity of solute release, R is the radius of reservoir, l is the thickness of coating, and V is the volume of a reservoir. We considered the reservoir to be a sphere, but the model can be easily modified for consideration of a reservoir of any geometry. The model is relevant to experimental data and predicts the general features of solute release from osmotic systems (Fig. 3.2):

—an increase in solute concentration in the core solution followed by acceleration of release;

—the steady state or maximum of release due to competition of processes of relaxation and the formation of saturated solution.

—decrease in the velocity of release followed by the tendency of activities of water in liquid phase and saturated solution to be the same (osmotic pressure difference tends to be zero).

We considered different concentration dependence of the diffusion coefficients of water $[D_w = D_{w0} \exp(k_{w1}C_w)]$ in attempt to predict possible kinetics of the solute release.

Curve 1 (Fig. 3.2) corresponds to the case of constant D_w and shows the above regularities. Curve 2 characterizes the case of exponential decrease of the diffusion coefficient ($k_{w1} < 0$) as the concentration of water increases. For this case the deceleration of the flux of solvent follows an increase in the activity of water in the saturated solution. Because of the equality of the fluxes of water and the dissolved solute, the velocity of solute release will be lower as time goes by. For this reason a polymer coating which is characterized by such a concentration dependence of the diffusion coefficient does not provide steady-state release.

A steady-state flux of the salt solution can be reached when the effect of dissolution is compensated by an increase in the diffusion coefficient of water based on the positive D_w (C_w) dependence. Curves 3 and 4 in Fig. 3.2 show the modeling of such a system. The positive dependence D_w (C_w) provides a steady-state release and the longer the k_{w1} the higher.

The application of this model to alteration of kinetics of release from osmotic polymer systems are considered in chapter 4.

Water uptake and transport parameters measured at 30°C for several available perfluorosulfonic acid membranes were compared in [16]. The

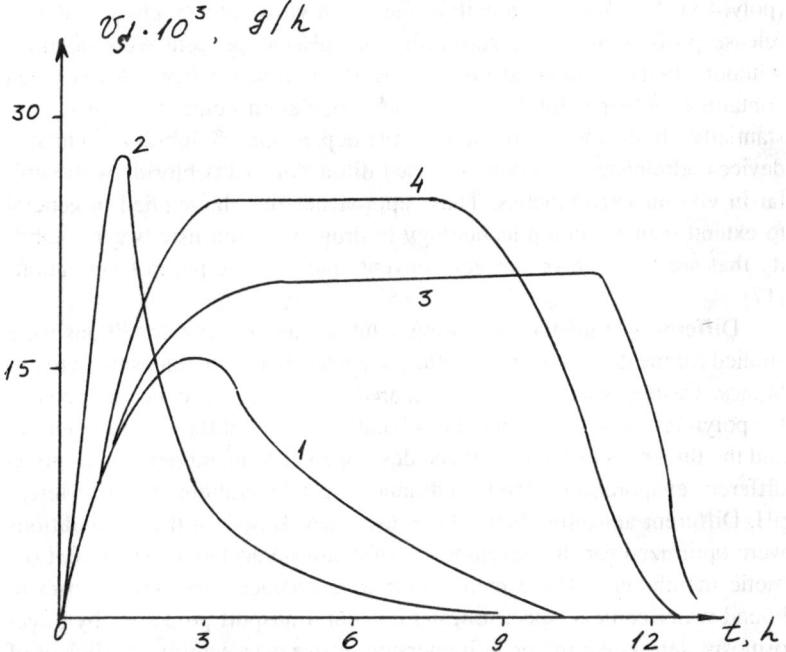

Figure 3.2. Computed kinetics for the controlled release of solutes from a reservoir system.
$D_{w0} = 10^{-8}$ cm^2/s, $C_{w0} = 0.01$ g/g.

water sorption characteristics, diffusion coefficient of water, electroosmotic drug, and protonic conductivity were determined for Nafion (R) 117, Membrane C, and Dow XUS 13204.10 developmental fuel cell membrane. The diffusion coefficient and conductivity of each of these membranes were determined as functions of membrane water content. Experimental determination of transport parameters, enabled Zawodzinski et al. to compare membranes without the skewing effects of extensive features such as membrane thickness which contributes in a nonlinear fashion to performance in polymer electrolyte fuel cells [16].

The drug release kinetics from controlled porosity osmotic pumps were effectively manipulated through application of either solubility- or resin-modulation methods [17]. The solubility of diltiazem hydrochloride was modulated (reduced) for an extended period of 12–14 hours through incorporation of controlled release sodium chloride elements into the core tablet formulations. Other diltiazem hydrochloride core tablets were prepared which contained the positively charged anion-exchange resin (poly(4-vinylpyridine)). In both instances, in vitro diltiazem hydrochloride release profiles that were zero-order and pH-independent were obtained without chemical modification of the drug. Release from devices that contained neither solubility- nor resin-modulation components was substantially first-order and highly pH-dependent. Solubility-modulated devices administered to dogs released diltiazem hydrochloride with similar in vivo/in vitro kinetics. These approaches may be applied in general to extend osmotic pump technology to drugs with intrinsic water solubility that are too high or low for conventional osmotic pump formulations [17].

Different parameters of casting solutions and casting conditions were studied for the development of cellulose acetate benzoate flat osmotic membranes. Casting solutions were prepared with different concentrations of the polymer, the additive, and the solvent; viscosity of the casting solution; and the thickness of the membrane developed. The membranes were given different evaporation periods and annealing temperatures under different pH. Different annealing baths were also used. Based on these, conditions were optimized for the development of cellulose acetate benzoate flat osmotic membranes. These membranes were characterized with respect to bound water content, specific water content, transport properties by direct osmosis, salt intake by direct immersion, water permeability coefficient of the dense membrane, diffusion coefficient, salt permeability, and salt distribution by electrical conductivity. Also, cellulose acetate benzoate membranes were compared with conventionally used cellulose acetate membranes [18].

A simple model was described for osmotic pumping drug release from membrane-coated formulations in [19]. The process requires at least two different areas in parallel in the membrane coat, which exhibit differing reflectivity to solute (drug and/or excipients). The model was used to calculate the reflectivity of membrane coats on potassium chloride tablets. Tablet coats of pure ethylcellulose (EC) had a low reflectivity of less than 0.005, and was not a major release rate regulating parameter. The higher reflectivity of the same membrane was 0.58, significantly lower than for a free relaxed film. The reduction in high reflectivity resulted in a corresponding lower drug release rate. Incorporation of hydroxypropyl methylcellulose (HPMC) in EC reduced the high reflectivity of the tablet coats even further. The reduction was accelerated when the HPMC content exceeds 24%. Since freezing water could be detected in free EC films with HPMC content of 24% and above, the formation of pores in these membrane coats was proposed. The pores were formed by the leaching of the water-soluble polymer HPMC, which was also found in the soaking liquid [19].

3.1.3. Transport controlled by biodegradation

The use of biodegradable polymers in controlled delivery devices has become increasingly attractive over the past years in medicine and agriculture for two main reasons:

—the polymer degrades in the body (or, for example in the soil) so there is no needs to remove the device after reservoir has been exhausted; and

—a zero-order release by surface erosion can be achieved when the polymer degrades.

One of the simplest way to use a biodegradable material is to encase the solute in a polymeric shell without chemical combination with the polymer [1]. We should note, however, that this way found less application than the biodegradable matrix system which will be considered below. For reservoir systems with biodegradable coatings the rate of release depends on the relation between the rate of solute diffusion across the shell and the rate of coating resolution. When steady-state flux is established, the rate of release (dm/dt) may be calculated as it is suggested in [20]

$$dm/dt = -DAdc/dl$$

where m is the mass of the solute, D is its diffusion coefficient in the biodegradable membrane, A is the surface area of the membrane, and l is its thickness.

The rate of erosion (ds/dt) is characterized by the constant K_{eff} and equal

$$ds/dt = K_{eff} A$$

The degradation rate decreases due to the reduction in the surface area, while the diffusion of solute accelerates due to an increase in the concentration gradient. In theory [21] it may be expected that the rate of release may pass maximum under certain conditions. The cases of the explosive release in such systems were recorded [22, 23] although generally it occurs in conformity with zero-order kinetics.

Biodegradable poly(d,l-lactide) (PDLLA) coated gentamicin/PDLLA and cefazolin/PDLLA cylinders were made for the controlled release of antibiotics [24]. The antibiotic release properties as well as release mechanisms (i.e., diffusion through channel, osmotic pressure, and polymer degradation) were investigated. Water soluble antibiotics could only be released through channels formed by connected drug particles and through polymer mass loss. Osmotic pressure played a key role by turning isolated drug clusters into connected channels through fracturing of the polymer matrix. The osmotic process of turning isolated clusters into connected clusters required time. This in turn gave a more gradual and sustained release than pure diffusion release through channel (i.e., without osmotic pressure effect involved). The effect of polymer biodegradation on release was significant when polymer mass loss started and at the same time there was a substantial amount of drug remaining in the device. In this case drug was released along with polymer mass loss. The cylinder core degraded faster than the cylinder shell when the longer gentamicin device was incubated in salt eluent. Gentamicin sulfate remained in the core and therefore catalyzed the polymer degradation. For the release into water or low osmotic eluent, three critical factors affected the release properties, namely, drug loading, drug particle size, and length of the coated cylinder. For antibiotic release from the coated cylinder with drug loading below the percolation threshold, as in the case of the 30% wt loaded gentamicin cylinder, the mechanism was a combination of pure diffusion through channels and osmotic pressure-induced diffusion through channels. Above the threshold (40–50% wt loaded gentamicin cylinders), the release was purely diffusion through channels and was very fast. A large size of drug particle resulted in a large degree of pure diffusion through channel at the same drug loading. The longer the cylinder, the longer and the slower the release. This gave a convenient method of being able to adjust the release properties [24].

Spray dried poly(lactide-co-glycolide) powders were used to prepare phenobarbitone matrix tablets [25]. The release of phenobarbitone was significantly sustained, indicating the suitability of using poly(lactide-co-glycolide) matrix tablets for long-term controlled drug delivery. Drug release profiles consisted of three regions: these were an initial region of relatively high release rate (burst effect), followed by an extended region of lower and essentially constant release rate from approx. 20 to 80% of drug release. This steady state release region was followed by a final one in which the rate of drug release fell off as exhaustion of the drug in the matrix approached. A composite mechanism of drug release is proposed involving diffusion of the drug through water-filled pores in the matrix and drug diffusion through the swollen polymer. The rate of drug release increased with an increase in the glycolide content of the polymers. Decreasing the polymer molecular weight caused initially an increase but later a decrease in the release rate. These results were discussed in [25] in terms of degradation, water uptake and swelling of the polymers upon immersion in aqueous media.

The above-mentioned (see section 1.1.1) TES-system has the time-controlled drug release property with a predesigned lag time. The drug release from the system may be initiated by destruction of the membrane. Ueda et al. studied the release model drug, metoprolol tartrate through biodegradable membranes in vivo. After five types of TES with different in vitro lag times were orally administered to dogs, plasma metoprolol concentration was monitored. There existed a good correlation between in vitro and in vivo lag time, while the extent of absorbed metoprolol decreased with prolongation of lag time. Next, the in vivo drug release behavior was directly investigated using five different colored TES with a lag time of two hours. Each TES was consecutively administered to the fasted dogs at predetermined intervals. The amount of metoprolol released was monitored by recovering the administered TES from the gastrointestinal tract. The in vivo release profile corresponded with the in vitro one. It is demonstrated that TES can release the drug in vivo conditions similarly to in vitro. Based on these results, the decrease of the absorption is suggested to be caused by increased hepatic first-pass metabolism of the drug due to the retarded release rate with longer lag time [26].

Marentette and Grosser proposed a model of a controlled-release drug delivery system and tested it by mathematical simulation [27]. The model consisted of a cylindrical drug reservoir coated with a permeable and erodible material. The release rate was relatively constant because, while the drug concentration was diminishing, the barrier was eroding. These factors might be balanced initially, but ultimately the release rate increased or decreased markedly depending on the permeability. Two kinds of permeability-dependent behaviors were observed: (i) for high-permeability coatings, the release rate was high and constant and then decreased as the drug was exhausted; and (ii) for low-permeability coatings, the release rate was low and constant and then increased as the barrier became thin. The proposed design combined two cylinders whose coatings had different permeability and exhibit complementary release behavior. Complete release occurred before total erosion of either coating. Optimization of the design and the design parameters led to a constant release rate (within 6%) until all of the drug was released from the device [27].

Svoboda et al. used fundamental microcapsule mass balance equations to describe the mass transport behavior of a diffusing species into or out of a microcapsule with a continuous, spherical core/shell structure. These equations demonstrate that at capsule shell thickness of 1/6, the capsule outer diameter maximizes the time required for the diffusing species to reach a specified concentration within the inner core region of the microcapsule. This optimum capsule thickness depends solely upon capsule geometry and corresponds to an active agent loading well below that often found in current microcapsules [28]. The theoretical analysis was applied to cellulose acetate phthalate (CAP) microcapsules which are used for enteric drug release. Unbuffered CAP solutions appeared unable to protect pH sensitive agents for the time required to reach the small intestine.

Buffered aqueous solutions were found to offer much greater protection than their unbuffered counterparts and appear capable of delivering pH sensitive agents to the small intestine [29].

Models of drug delivery devices that employ erodible permeable coatings must take care to avoid the unacceptably high rate of release that arises as the erodible coating disappears and the barrier to drug release vanishes. One solution to this safety problem has been to exhaust the drug reservoir just before this condition occurs. This design has the disadvantage of placing demands of high accuracy on the quality control in the fabrication of the device. A drug delivery system of cylindrical symmetry was proposed that uses two permeable coatings on a drug-containing core. Only the outer of the two coatings is erodible; the inner safety coating and the core are inert. Calculations were performed to design the device that can yield constant drug delivery rates while avoiding the possibility of explosive late drug release [30].

3.2. MONOLITHIC DEVICES

3.2.1. Diffusion controlled transport

The monolithic (or matrix) devices assume the load of solute in polymer matrix either in the form of dispersion or in solution with the consequent removal of the solvent [31]. This provides a wide variety in the mechanism of release, its kinetics, and the rate. As pointed out in the previous chapter, the diffusion of solute may be affected by mechanical and structural properties of the polymer, the solubility of water and solute, and other factors. Accordingly, the kinetics of the release may be described by the "differential swelling stress model" (Eqs. 2.3–2.11) or Higuchi dependence

$$C_s/C_{s0} = 2(D_{sF}t/\pi l^2)^{1/2}$$

or any other relevant model reported in chapter 2.

The kinetics of release also depends on the conditions of application of the particular device. For example, the solution of general equation (3.1) (which is the same for reservoir and matrix devices) for a semi-infinite medium takes the form

$$M_s = C_{s0}A[erf (X/4D_st)^{1/2}]^{-1}(D_st/\pi)^{1/2} \qquad (3.14)$$

where X is the coordinate of the moving front of the solute which can be found from Eq. 3.15, proposed by Lee [32]

$$\pi^{1/2}d_x \exp(X^2/D_st)\mathrm{erf}\,[(X/2D_st)^{1/2}] = S_s/(C_{s0} - S_s) \qquad (3.15)$$

As before D_s and S_s are the diffusion coefficient and the solubility of solute, C_{s0} is its initial concentration in the matrix, A is the surface area, and M_s is overall quantity of the compound released at time t.

The generalized diffusional equation for a polymer matrix of any geometric form is given in [1] by

$$\delta C_s/\delta t = r^{1-n}\,\delta/\delta r[D_s\,r^{n-1}\,\delta C_s/\delta r] \qquad (3.16)$$

where r is the generalized coordinate of diffusion, and n is the generalized order (n = 1 for a plate, n = 2 for a cylinder, and n = 3 for a sphere). If the initial distribution of a solute in a matrix is given by a function f(z), then the solution of Eq. 3.16 takes, for example, for a plate the form [32]

$$C_s^{\infty}(z,\tau) = 2\sum_{k=0}^{\infty} \exp\,[-(2k+1)^2\,\pi^2\tau/4 - \cos\,[(k+1/2)\pi z]\,I \qquad (3.17)$$

where $I = \int f(z)\cos\,[(k+1/2)\pi z\,dz;\ z = 2r/l$, and $t = 4D_st/l^2$.

Accordingly the overall quantity of the released solute is calculated as

$$M_{st}/M_{s\infty} = 1 - (\Sigma \leqslant -1)^{k+1}(2k+1)^{-1}I\,\exp\,[-(2k+1)^2)\pi^2\tau/4])\Big/$$
$$(\Sigma(-1)^{k+1}\,(2k+1)^{-1}\,I).$$

Lee derived the corresponding expressions for a cylinder and sphere [32].

Thus, the initial distribution of the solute in polymer matrix must have an effect on the kinetics of its desorption. Lee reported the numerical analysis of this effect in [32]. He referred the results of his computations to the behavior of several monolithic drug delivery systems. The same approach found application also in the development of anticorrosion [1, 33] and agricultural [34] polymer systems.

The mechanisms governing the release of osmotically active agents from hydrophobic polymeric monoliths were outlined in [35]. The release kinetics were loading dependent. Below the percolation threshold, the bulk of the loaded excipient was released with zero order kinetics. Above the percolation threshold, diffusional and osmotic release mechanisms occurred simultaneously resulting in a release profile that appeared to be diffusionally controlled but whose slope increased with excipient osmotic activity. Amsden and coauthors developed a model to overcome some usual limitations. The main parameters affecting release were determined to be the osmotic activity, saturation concentration and density of the incorporated agent, the tensile strength, elastic modulus and hydraulic permeability of the polymer, and the fraction of particles released by dissolution and diffusion. The model was correlated against data from experiments and the literature and provided a reasonable fit [35].

Ethylene-co-vinyl acetate monoliths containing osmotic excipients of varying particle sizes and at various volume fractions were reported by Amsden and Cheng in [36]. The excipients chosen were potassium chloride, sodium chloride, sodium salicylate, sodium phosphate monobasic, sodium carbonate, and bovine serum albumin. The total fraction releasable was found to increase as the volumetric loading and the excipient osmotic activity increased, and was weakly inversely dependent on particle size. The total fraction released was enhanced by osmotic pressure induced polymer rupturing of encapsulated particles. A model was developed, based on the concept of a distribution of particle wall thickness around encapsulated particles in the monolith, and the existence of a critical wall thickness for rupture, to explain the fraction releasable enhancement. The model effectively explained the contribution of osmotic rupturing to the overall fraction releasable [36].

Jenuqin et al. investigated how the rate of drug release from a polymeric matrix system was influenced by the physical and chemical properties of the monolithic films [37]. The model drugs, salicylic acid and chlorpheniramine maleate, and two poly(methyl methacrylate) copolymers of different permeability (Eudragit RL and Eudragit RS), with and without additional adjuvants, were used to form monolithic matrix films for controlled drug release. Adjuvants, including polyethylene glycols (PEG 400 and PEG 8000) and poly(vinylpyrrolidones) (PVP-K15 and PVP-K90), were incorporated into films of Eudragit RL PM and Eudragit RS PM. The moisture permeation constant, glass transition temperature T_g, tensile strength, and drug release profiles were determined for each acrylic resin slab to correlate the physicochemical and physicomechanical film properties to observed drug release. Faster rates of drug diffusion were observed with the addition of PEG 400 to the films, because of its plasticizing effect and the resultant increased moisture permeability of the matrix. An exception existed with the Eudragit RL PM film containing salicylic acid where drug-polymer interactions inhibited drug diffusion. The small changes in moisture permeability, T_g, and tensile strength observed with incorporation of the PVPs had an insignificant influence on the dissolution results for salicylic acid from Eudragit RS PM films. Increases in the tensile strength and T_g after addition of PVP to the Eudragit RS PM matrix support the observed decreased rate of diffusion for chlorpheniramine maleate. The pores formed by migration of the hydrophilic adjuvants from the films altered the diffusion kinetics of the matrix, compared with that of the nonporous polymer, when only the antihistamine was present [37].

Prolonged-release spherical micro-matrices of ibuprofen with Eudragit RS were prepared using a novel emulsion-solvent diffusion method. Those particles were termed "microsponges" due to their characteristic sponge-like texture and unique dissolution and compression properties, unlike conventional microcapsules or microspheres. The internal porosity of microsponges could be easily controlled by changing the concentration of the drug and the polymer in the emulsion droplet (ethanol). With lower concentration of ibuprofen in the ethanol, the resultant microsponges had a higher porosity, about 50%. The drug release rate from the microsponges was interpreted by the Higuchi model of spherical matrices, which depended only on their internal porosity of the microsponges when size distribution and drug content were the same. The tortuosity in the microsponges was found to be almost constant (3–4) irrespective of porosity, suggesting the same internal texture. Microsponge compressibility was much improved over the physical mixture of the drug and polymer owing to the plastic deformation of their sponge-like structure. The more porous microsponges produced stronger tablets [38].

Solved mixtures of hydrogels (SMH), composed of hydroxypropyl-cellulose (HPC), a pseudo-hydrogel, and ethylcellulose, a water-insoluble polymer, were prepared by solvent evaporation. Phenylpropanolamine hydrochloride was used as a model drug. The amount of drug released from a matrix tablet containing SMH powder and the drug increased with decreasing weight fraction of HPC (WFH) in SMH. SMH showed improved properties compared to HPC alone, for example, flowability and hydroscopicity. These properties increased with decreasing WFH. The sorption apparatus was capable of an immediate, sensitive, and accurate response to initial moisture sorption. [39].

Conventional monolithic drug delivery devices in which drug is uniformly dispersed typically show Fickian release rates. To achieve zero-order release, monolithic devices of hydrophilic polymer were prepared, loaded with hydrophobic drug [40]. Hydrophobic indomethacin was loaded into hydrophilic matrices of poly(N-isopropyl acrylamide (IPAAm)-co-alkyl methacrylate (RMA)) whose hydrophilicity can be varied by temperature without changing the chemical structure. Drug release experiments were performed in phosphate buffered saline. Under conditions of high drug loading in highly hydrophilic polymeric matrices, the release patterns were observed to be nearly zero-order. This result can be explained in terms of increasing diffusivity in polymeric matrices after drug is released, because loaded hydrophobic drug suppressed the swelling of polymeric matrices. An equation predicting drug release based on diffusivity changes of the polymeric matrices was derived, and a good agreement was found between the experimental results and the theoretical release simulation [40].

A mathematical model was presented in [41] for the description of transdermal drug delivery from a matrix-type delivery device. The model is partly diffusional and partly compartmental in nature. The matrix and stratum corneum are both considered to be diffusion layers, connected to a three-compartment model representing the viable epidermis/dermis, plasma, and peripheral tissues. The diffusion equation was solved numerically for the two diffusion layers under nonsink conditions. The ordinary differential equations for the compartmental model were also solved numerically. Combination of the two numerical solutions yielded a model which directly related the properties of the matrix to the profile of drug mass in the plasma and the urinary excretion profile. The model was first used to analyze data obtained from an in vivo trial of a matrix-type transdermal delivery device for the drug clenbuterol. Fitting of the model to the profile of drug concentration in the plasma, the urinary excretion profile, and the mass of drug remaining in the matrix with a modified simplex method yielded values for the model constants. These compared very favorably with independent values taken from the literature. Simulations of the influences of drug diffusivity within the stratum corneum, drug loading in the matrix, matrix thickness, and drug diffusivity within the matrix on the profile of drug concentration in the plasma were then made. The model was not restricted to a steady state nor does it specify particular drug release kinetics from the matrix. It did assume isotropic diffusion layers and spontaneous partitioning at boundaries [41].

Vanbommel et al. [42] presented a mathematical model with which acetaminophen release curves from the gradient matrix system (GMS) with different geometry (slabs and spheres) can be described. Diffusion was considered the rate-controlling step in the release process. Position- and time-dependent diffusion coefficients accounted for the changes in the matrix structure due to the initial different loading concentrations and due to changes during the release process. Release of acetaminophen from slab model systems and from spherical systems was adequately explained. Release curves of another model drug, mebeverine HCl, from planar GMS formulations were adequately predicted by the same model [42].

The release of bovine serum albumin (BSA) from two types of ethylene-co-vinyl acetate (EVAc) polymer matrices was studied over the temperature range 4–50°C. Protein release and weight change of the matrices were evaluated in vitro. The copolymers were characterized using differential scanning calorimetry (DSC) and thermomechanical analysis (TMA). During release, the devices initially exhibited a rapid increase in weight to a maximum, followed by a more gradual decrease for the duration of the release. The time to the maximum weight and the magnitude of the maximum weight gain were temperature dependent. These effects were related to the temperature-dependent diffusivity of the BSA and elastic modulus of the EVAc. The DSC and TMA reveal melting of the crystalline phase of the polymer. The corresponding loss of mechanical integrity of the polymer leads to anomalous weight gains at these temperatures. The observed swelling and release were explained by a model in which the osmotic pressure of the protein within the pore network caused elastic deformation of the polymer matrix [43].

A heterocyclic methacrylate polymer system, developed originally as a low shrinkage polymer system, was investigated as a drug release polymer and as a biomaterial for encouraging bone or cartilage regeneration. The system was based on poly(ethyl methacrylate) polymer powder mixed with tetrahydrofurfuryl methacrylate monomer and polymerized at room temperature (PEM/THFM). Promising results were obtained with this biomaterial, and hence its water uptake properties were investigated in detail, in order to throw some light on the release processes that are involved in vivo and in vitro. Water soluble large molecule analogues were incorporated into the system; these additives increased the water uptake of the system. Isobornyl methacrylate was used as a diluent for the monomer to further reduce the water uptake of the system. In all cases the uptake kinetics did not obey simple diffusion theory, the process being very prolonged and complex [44].

Smart investigated the development of monolithic matrices with controlled release and mucosa-adhesive properties. After an initial screening procedure for formulations that showed stability and minimal swelling, the rate of release of a model water soluble drug from various polyacrylic acid containing matrices was evaluated. All the formulations gave a prolonged drug release relative to a lactose containing control formulation. A formulation containing Carbopol 934P and CaCl2 was found to give the slowest rate of drug release (t50% of 7.77h), with release kinetics nearest to the ideal zero order. When tested in a modified tensiometer it was found that the inclusion of a relatively high loading of a model drug did not adversely affect the adhesive properties of these formulations [45].

3.2.2. Transport controlled by biodegradation

The kinetic study of the release of low-molecular weight solutes (LMS) from the biodegradable polymer monoliths requires the analysis of the possible LMS-polymer combinations. For example, almost all the polymeric drugs may be divided into several groups which are presented in Table 3.1 [1].

Thus, there are three types of combination of LMS molecules with the polymer matrix which are characterized by different release dynamics [1]:

Table 3.1. Classification of polymeric drugs.

Medical form	Features	Method of administration
Polymer solution	Polymer is the solvent	Injection
	Drug is chemically combined	Ophthalmic drops
Polymer gel, ointment	Polymer is the filler or adhesive	Orally, externally
suppository	Drug is chemically combined with the polymer molecule	Orally, externally
Synthetic sponge or film	Drug is distributed over sponge or film	Externally
	Drug is chemically combined with the polymer molecule	Externally
Microencapsulated drugs	Shell of bioinert polymer	Injections, suspension, orally
	Shell of biodegradable polymer	Injections, suspension, orally
Drug-bearing polymer	Bioinert polymers	Hypodermically
implant	Biodegradable polymers	Externally
Ophthalmic film	On the basis of resolvable polymer	

—the drug molecules are included in the main chain of the polymer;

—covalent combination of the drug to the side chains of the polymer chain of the polymer (random or regularly); and

—the drug is distributed in polymer solution or in solid polymer monolith without chemical bonds.

There are certain difficulties with including the LMS molecule in the main chain of the polymer. For this reason a limited number of such systems were reported so far. One example was polyamide prepared by polycondensation of δ-amino-levulinic acid chloroanhydride [46]. Since most polymers of this type are heterochain compounds they undergo the hydrolysis in the body [47, 48] and, usually the hydrolysis is random [49–53] rather than a depolymerization [54–56]. In this case, LMS is not released with each bond scission, and the number of effective scissions is higher the lower the molecular weight of the polymer [57].

The first state of release is described by the following equation

$$C_{LMS} = [LMS]_0 [1 - \exp(-kt)][2 + (P_n - 2)(1 - \exp(-kt)]$$

where k is the hydrolytic degradation rate constant, P_n is the degree of polymerization, $[LMS]_0$ is the limiting LMS concentration arising in the case of simultaneous opening of all the bonds of the main chain, and t is the time as always. As LMS accumulates, for example in the blood, its concentration passes a maximum and then decreases due to elimination through metabolism and excretion.

If LMS molecules are attached to side chains of dissolved polymer, one can apply the mathematical approach reported in [1, 20]. Ignoring the polymer elimination rate, there was obtained the equation for kinetics of LMS accumulation in blood

$$d[LMS]/dt = k_h\{[LMS]_0 - [LMS] - k_{el}[LMS]^2\}$$

where k_h is the effective rate constant of LMS release and k_{el} is the elimination rate constant.

The rate of release is maximal when

$$[LMS]_{max} = k_h[LMS]_0/(k_h - k_{el})$$

and half life-time of the device is given by

$$\tau_{1/2} = 1/(k_h - k_{el}).$$

If there are no chemical bonds between the solute and polymer matrix, a mathematical description of an erosive monolithic device should take into account (i) solvent and biological medium diffusion into polymer, and (ii) erosion front movement leading to the weight loss and the reduction of size of the polymer [1]. Lee [58] proposed the generalized scheme of the solute distribution in erosive polymer which takes into account the alteration of the geometry of the sample

$$C(l[t], t) = kc_V$$

$$C(R[t],t) = C_s^T$$

$$D_s \delta C_s/\delta x = (C_s^0 - C_s^T)\delta R/\delta t \qquad x = R(t)$$

$$D_s \delta C_s/\delta t = -\lambda l_0(n+1)^{-1}\delta C_s/\delta t \qquad x = l(t)$$

$$R(0) = l(0) = l_0$$

where λ is the ratio of volumes of solvent (v) and polymer, C_s^0 and C_s^T are initial concentration of solute and its thermodynamic solubility in the polymer, respectively.

The analysis of the effect of the initial distribution of solute on the kinetics of its release was made by Lee in [32] where he made the following important theoretical conclusions:

1. For a film with uniform distribution of the solute the constant rate of release may be expected. Such a situation never occurs for a sphere or cylinder.
2. The parabolic distribution of solute in the matrix provides the constant rate of release for a plate, sphere, and cylinder.
3. The S-shape distribution leads to a short time constant rate of release for a sphere or cylinder.
4. The pulsed kinetics of release is the consequence of the stepped initial distribution of a solute.

All examples of the above distributions are shown in Fig. 3.3

Joshi and Himmelstein proposed a mathematical model for the analysis of the basic physicochemical determinants that yields experimentally verifiable predictions of controlled release of bioactive agents from eroding polymeric matrices [59]. Their model was applied to the erosion characteristics of acid catalyzed erosion of poly (ortho ester) since there was considerable information in the literature on the detailed physical and chemical performance of this system. The analysis showed that the dynamic changes in polymer matrix properties (namely, simultaneous reaction-diffusion-transport of matrix constituents, moving diffusion and water fronts, water-polymer partition coefficients, solubility of water and diffusivity of matrix components as a function of the extent of acid catalyzed matrix hydrolysis) play a significant role in regulating the release kinetics of bioactive agents. The analysis of poly (ortho ester) erosion as a test system predicts experimentally verifiable measurable quantities: release characteristics of the incorporated bioactive agent, water penetration into the matrix, catalysis by the anhydride, and catalytic degradation of the polymer matrix. Further, an estimate of the concentration of unbroken polymer backbone linkages and hence the molecular weight of the polymer disc with time was obtained using random scission kinetics. The Thiele module was a good indicator of surface versus bulk erosion and has been successfully applied to characterize the erosion characteristics of the poly (ortho ester) system. Finally, an analysis for the basic design, interpretation, and prediction of overall release modes (bulk versus surface erosion) from devices that rely on simultaneous reaction and diffusion controlled erosion phenomenon was presented [59].

A drug delivery system for biologically active agents targeted to specific cells can be used to improve tissue repair in orthopedics. The system should be controllable and capable of drug release over an extended period of time. Biodegradable, membrane-moderated, monolithic micro-

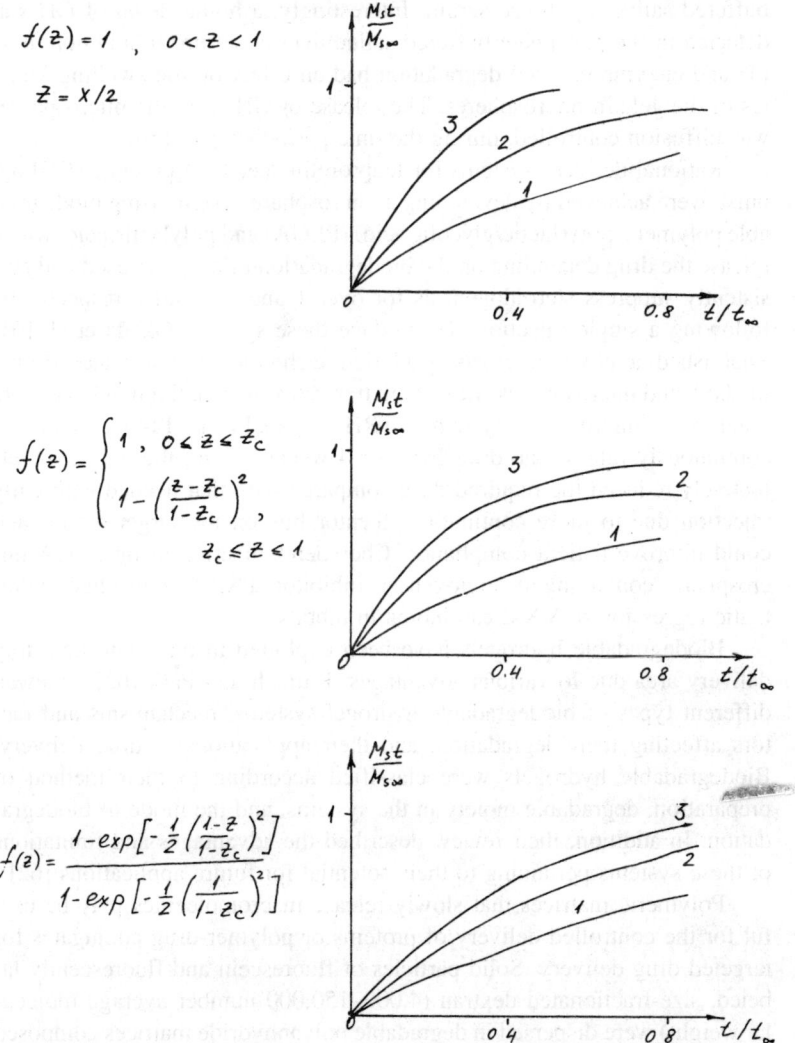

Figure 3.3. Effect of initial distribution of the drug in a polymer on the kinetics of release for a plane sheet (a), a sphere (b), and a cylinder (c). z_c is the critical point of inflection, and z is the relative coordinate of diffusion.

spheres for the controlled release of growth hormone (GH) were developed, and the release of GH was monitored in vitro. Cross-linked gelatin microspheres were used as the vehicle, with the drug dispersed within the gelatin. The amount of GH released from the microspheres was increased following ultrasonication. The release of GH was monitored in phosphate buffered saline and horse serum. Interestingly, a higher level of GH was detected in the phosphate buffered saline than in serum. In addition, both pH and enzyme-induced degradation had an effect on the swelling kinetics of the gelatin microspheres. The release of GH from the microspheres was diffusion controlled, during the time period studied [60].

Rational delivery systems for leuprorelin acetate, a potent LHRH agonist, were achieved by developing a microsphere system using biodegradable polymers, poly(lactic/glycolic acid) (PLGA) and polylactic acid, which release the drug depending on the biodegradation of polymer used and persistently suppress steroidogenesis for over 1 and 3 months, respectively, following a single injection. To produce these systems Okada et al. [61] established a novel microencapsulation technique, the in-water drying method, and microspheres with a high trap ratio and small initial burst were obtained. A microsphere system of TRH prepared using PLGA could also continuously release the drug for 2 or 4 weeks. Using these systems effectively reduced the required dose compared with that needed with daily injection due to more continuous receptor hits on the target organs and could improve patient compliance. Chemoembolization using PLGA microspheres containing an angiogenesis inhibitor, TNP-470, resulted in dramatic regression of VX-2 carcinoma in rabbits.

Biodegradable hydrogels have been exploited in the controlled drug delivery area due to various advantages. Kamath and Park [62] reviewed different types of biodegradable hydrogel systems, mechanisms and factors affecting their degradation, and their applications in drug delivery. Biodegradable hydrogels were classified according to their method of preparation, degradable moiety in the systems, and the mode of biodegradation. In addition, their review described the advantages and limitations of these systems pertaining to their potential for future applications [62].

Polymeric matrices that slowly release macromolecules may be useful for the controlled delivery of proteins or polymer-drug conjugates for targeted drug delivery. Solid particles of fluorescein and fluorescently labeled, size-fractionated dextran (4,000–150,000 number average molecular weight) were dispersed in degradable polyanhydride matrices composed of a 1:1 copolymer of fatty acid dimers and sebacic acid. The release of macromolecules from the polymer matrix into buffered saline was measured; changes in the polymer during immersion were monitored by infrared spectroscopy, differential scanning calorimetry, and scanning electron microscopy. Although significant hydrolysis of the polymer occurred within the first day, the matrices remained intact and water-soluble tracers were slowly released for several days. During polymer hydrolysis and erosion, micron-sized pores developed throughout the 2 mm thick polymer matrix, permitting water penetration into the matrix and tracer diffusion out of the matrix. The rate of tracer release from the matrix depended on tracer particle size; rates of fluorescein isothiocyanate dextran release were controlled by adjusting the size of particles dispersed in the matrix [63].

Diltiazem release from drug delivery systems, both microparticulate and pelleted, prepared from poly-lactide-co-glycolide co-evaporates is controlled by co-polymer decomposition. In contrast, drug release from systems prepared by direct compression of drug/co-polymer mixtures more closely approximates matrix type diffusion. Significant differences in water uptake and solid state form of the drug were observed depending on the method of preparation. Drug release from co-evaporate systems was fitted to a model which reflects the formation, spread, and termination of activated nuclei of decomposition. Model parameters representing the induction and acceleration of decomposition, may prove useful in correlating release with the physicochemical properties of the drug and polymer, ultimately enabling a more quantitative approach to the design of biodegradable poly-α-hydroxy aliphatic ester based drug delivery systems [64].

The mechanism of release of drugs from poly(glycolic acid-co-DL-lactic acid) (PGLA) was investigated in [65]. Films of a 1:1 copolymer were either solvent-cast or prepared by compression of micronized powders. The water content, glass transition temperature, hydrolytic chain scission, and weight loss of PGLA were measured as a function of time in deionized (DI) water and phosphate-buffered saline (PBS) at pH = 7.4 and 37°C. Changes in these properties were compared with the rates of release of testosterone and bovine serum albumin (BSA) from PGLA films in order to establish their contribution to the delivery mechanism. Rates of release were initially slow, but exhibited a characteristic acceleration between 10 and 15 days. The rates were largely independent of the medium, DI-water or PBS, and the method of device fabrication. There was no correlation of the change in drug release kinetics with the glass transition temperature, which in water decreased from 45°C (dry state) to 30°C within 1 hour and to 10°C after 2 weeks. The water content of PGLA increased with time and depended on the medium; values greater than 20% were the result of pores formed by polymer dissolution within the polymer bulk. The decrease in the molecular weight with time exhibited the semi-logarithmic relationship characteristic of the hydrolytic chain scission of other aliphatic polyesters. Polymer weight loss was not observed until the M_n had decreased to several hundred, and coincided with the onset of more rapid drug release. Microscopy suggested that polymer weight loss was a bulk process. This was confirmed by the rate of drug release, which was identical to the rate of polymer weight loss but independent of its surface-to-volume ratio [65].

A novel drug delivery system was developed using a monoglyceride (Glycerol Monostearate) and a water-soluble release rate modifier as the matrix. Cefuroxime sodium (Zinacef-R) was chosen as a model drug in this study. Formulations (cylindrical implants 6×6 mm) were prepared by a melt-dispersion method. Dissolution studies were performed using USP paddle method. The effect of glycerol, PEG 400, and their combination on drug release profiles was studied. Two assay methods (UV and HPLC) for cefuroxime analysis were compared. Percent recovery from four formulations (A–D) was higher with UV than HPLC assay. While both UV and HPLC assay methods were developed for cefuroxime, only HPLC assay is stability indicating. Glycerol showed a higher accelerating effect than PEG 400 on the drug release. All formulations exhibited extended release of cefuroxime. Degradation of cefuroxime occurred mainly during dissolution suggesting drug stability in the formulations [66].

Five polyphosphazenes with amino acid esters and imidazole as substituents at the phosphorus atom were recently synthesized: POF I with AlaOEt as only substituent, POF II and POF III with AlaOEt/Imidazole 90/10 and 70/30 molar ratios, POF IV and POF V with PheOEt/Imidazole 90/10 and 70/30 molar ratios, respectively. The polymers were characterized by H-1 NMR spectrometry, viscosimetry, swelling, and water degradability. Matrices were prepared in the form of films by solvent evaporation and spheres by solvent extraction. The release of model drugs naproxen, narciclasine and acetyl tryptophanamide from these polymers was evaluated by Calicati et al. [67]. The films were found more suitable for drug release, which was observed to be related to matrix degradability. Polymers II and III, degrading more rapidly, release drugs at the highest rates, whereas polymers I and IV degrade and release drugs more slowly.

Poly(DL-lactic acid), synthesized by Lalla and Sapna from DL-lactic acid, was used to prepare microspheres containing piroxicam, using a solvent evaporation technique. The microspheres obtained were characterized for their surface characteristics (by SEM), surface charge, density, particle size distribution, glass transition temperature, drug incorporation and encapsulation efficiency, IR spectroscopy, and in vitro drug release. The suspension of microspheres was evaluated for its syringeability. The effect of channeling agents such as PVP and PEG 6000 on in vitro drug release was studied. The effect of gamma-radiation on poly(DL-lactic acid) and on the in vitro release of piroxicam from the microspheres was also reported in [68].

Sah et al. [69] prepared biodegradable microcapsules from various polymer compositions such as poly-D,L-lactic acid or poly-L-lactic acid and poly-D,L-lactide-co-glycolide for the controlled release of a model protein bovine serum albumin (BSA). ELISA data indicated that the binding activity of BSA toward anti-BSA antibody was not lost during the microencapsulation process and the release from microcapsules. The effect of different microcapsule formulations on release profiles of the protein was investigated. It was found that the polymer composition, total amount of polymers, and protein loading were factors involved in the determination of release profiles of BSA. The degradation of various microcapsules was also investigated. Factors such as degrees of the hydrolytic chain scission of ester linkages, water content, and internal structures were compared. The microcapsule formulation influenced the porosity and the degree of concentration gradient of BSA. These were related to the initial burst effects and release patterns. All these factors interplayed their roles on the release characteristics of BSA. Depending on the microcapsule formulation, controllable and predictable release profiles of BSA that could be described by either zero- or first-order kinetics were observed [69].

More efficient means of administering drugs at controlled rates are continuously being sought. An important alternative is biodegradable polymers which do not have to be removed once implanted. A class of biodegradable polymers studied by several groups is the polyanhydrides. The synthesis of a novel class of polyanhydrides containing skeletal amino acid residues was recently described [70]. The degradation kinetics and the potential applications in controlled drug release of one of these polymers, poly(N-trimellitylimido-beta-alanine-co-sebacic anhydride) (PTIASA; 20:80), were investigated simultaneously for the first time by Coebas et al. [71]. Drug-polymer discs were prepared by compression molding using either acid orange or sodium salicylate as the model drug (approximately 30% weight loading). In vitro incubation experiments were carried out in buffer solutions of pH = 7.4 and pH = 10.0 at 37°C. The appearance of beta-alanine (a polymer by-product) and drug in solution was followed spectrophotometrically for up to 240 hours. Mass balances were then performed to determine the rates of polymer degradation and drug release. Drug-free PTIASA discs decreased in thickness throughout the incubation period at both pHs but maintained their shape. The degradation rates (in terms of the appearance of beta-alanine in solution) were relatively constant (zero-order kinetics) and higher at pH = 10.0. The drug-free discs degraded completely in 7–8 days at pH 10.0 and in about 10 days at pH 7.4. On the other hand, the drug-loaded PTIASA discs swelled initially and disintegrated partially later on. In the acid orange-PTIASA experiments about 90% of the drug was released in the first 15 hours of incubation, the remaining 10% being released at a relatively constant and slower rate thereafter. When sodium salicylate was used as the test drug, the magnitude of the initial burst in drug release was reduced and the useful life of the formulations extended. This study clearly showed the potential use of these novel amino acid-containing polyanhydrides in controlled release formulations.

An erodible association polymer system based on blends of cellulose acetate phthalate (CAP) and Pluronic F127, a block copolymer of poly(ethylene oxide) and poly(propylene oxide) was investigated by Xu and Lee [72] for its applicability to rate-programmed drug delivery. The compatibility and thermal properties were characterized by DSC and FTIR. Results from the thermal analysis indicated that the blends were compatible above 50% CAP, as revealed by a single composition-dependent glass transition temperature. The existence of molecular association through intermolecular hydrogen bonding between the carboxylic acid and the ether oxygen groups was supported by the observation of an upward shift in the IR carbonyl stretching frequency at increasing Pluronic F127 concentrations. Using theophylline as a model drug, the in vitro polymer erosion and drug release characteristics of the present polymer system were evaluated at different buffer pHs on a rotating-disk apparatus. The results showed that the rates of both polymer erosion and drug release increase with the Pluronic F127 concentration in the blend. Further, at pH = 4, the polymer erosion was minimal and the theophylline release appeared to be governed mainly by diffusion through the polymer matrix. In contrast, at pH = 7.4, the theophylline release was controlled primarily by the polymer surface erosion. To demonstrate the unique approach to programmed drug release based on the concept of nonuniform initial drug distribution, pulsatile patterns of drug release were achieved successfully from the surface-erodible polymer system using a multilaminate sample design with alternating drug-loaded layers. The results suggested that the pulsing frequency and peak rate of such pulsatile drug delivery are pH dependent; however, they can be modulated by varying the thickness, drug loading, and erosion rate of the constituent layers in the multilaminate [72].

Aso and coauthors [73] investigated drug release and matrix degradation of poly (D,L-lactide) microspheres with different glass transition temperatures (T_g) at various temperatures in order to clarify the effect of temperature on mechanisms of drug release and matrix degradation. At temperatures above T_g, the average molecular weight of the polymer decreased markedly during drug release. Progesterone release was faster than microsphere weight loss, and could be fitted to the Higuchi equation. These results suggest that diffusion from the bulk of the matrix contributed to drug release at temperatures above T_g. In contrast, at temperatures below the T_g of the microspheres, the average molecular weight of the polymer did not change throughout the experimental period and matrix degradation was restricted to the matrix surface. Release of progesterone was due mainly to surface erosion. These results suggest that, even in the case of polylactide, drug release can be controlled only by surface erosion [73].

Recently, Fischel-Ghodsian and Newton mathematically modeled and simulated the drug release kinetics from a degradable polymer undergoing surface erosion [74]. Their approach considers drug particles homogeneously dispersed in the matrix, and cases of both a single polymer matrix and a matrix surrounded by a membrane were analyzed. A schematic representation of their model for a polymer without and with a membrane is shown in Fig. 3.4. The parameter "a" represents the full or half thickness of the device, and other surfaces are assumed to be coated so that the surface area remains constant during drug release. C_0 is the initial drug loading, C_s is the drug solubility in aqueous media, C_p and C_m are concentrations of the drug in the polymer and in the membrane on their interface. Diffusion and surface erosion are considered by two moving boundaries u(t) and s(t), respectively. As the matrix is exposed to the release media and the surface layer becomes depleted of the drug, so the diffusion front is shown to pass the eroding front.

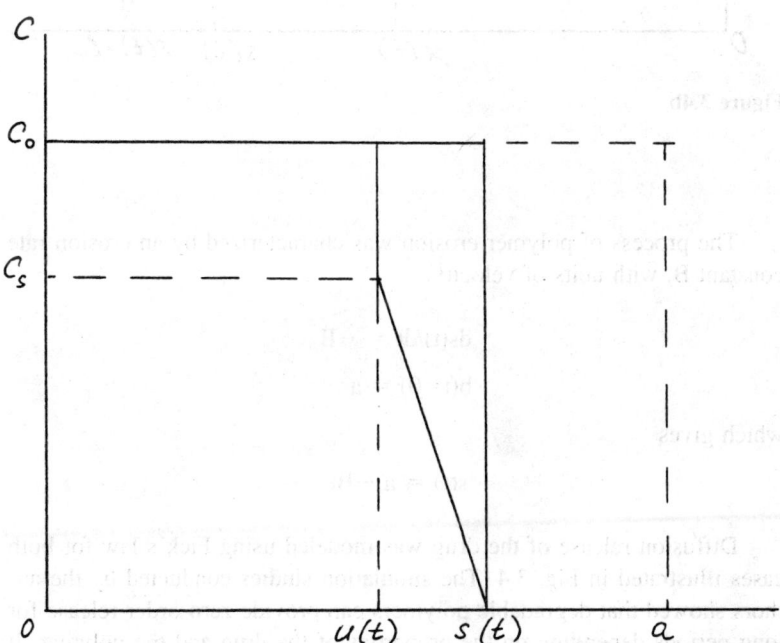

Figure 3.4a. Schematic representation of the model for drug release from a degradable polymer without (a) and with (b) a membrane.

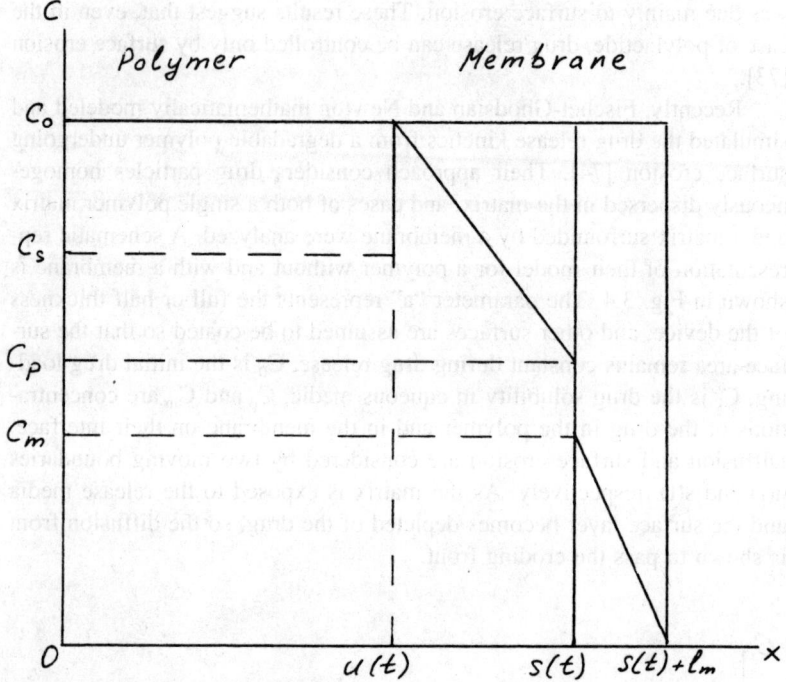

Figure 3.4b

The process of polymer erosion was characterized by an erosion rate constant B, with units of velocity:

$$ds(t)/dt = -B$$

$$b(t=0) = a$$

which gives

$$s(t) = a - Bt.$$

Diffusion release of the drug was modeled using Fick's law for both cases illustrated in Fig. 3.4. The simulation studies conducted by the authors showed that degradable polymers can provide zero-order release for long periods depending on the properties of the drug and the polymer. It was concluded that:

1. The period needed to achieve zero-order release decreases as the erosion rate constant increases and the drug diffusion coefficient decreases. When the polymer is surrounded by a membrane, this period increases and total diffusion lenth (l) increases.
2. A membrane extends both the lifetime of the device and the period over which the drug is released by zero-order kinetics.
3. There is a range where the device does not show zero-order kinetics since the period needed to achieve a zero-order release exceeds the lifetime of the device.
4. At steady state, the release rate of the drug is proportional to polymer erosion rate, surface area of the device, and drug loading, and is not affected by the membrane.

The above conclusions provided a useful theoretical basis for the design of degradable polymeric devices for the delivery of therapeutic agents and other substances.

REFERENCES

[1] A.L. Iordanskii, T.E. Rudakova, G.E. Zaikov. *Interaction of Polymers with Bioactive and Corrosive Media.* VSP, Utrecht, The Netherlands (1994).

[2] T. Roseman, L.J. Larion, S.S. Batler. *J. Pharm. Sci., 70,* 562 (1981).

[3] R.Yu. Kosenko, A.L. Iordanskii, G.E. Zaikov. *Khimiko-Farmatsevticheskii Zhurnal (Chem. Pharm. J.),* 23, 741 (1989) (In Russian).

[4] G.B. Vandenberg, C.A. Smilders. *J. Membr. Sci., 73,* 103 (1992).

[5] P.I. Lee. *J. Pharm. Sci., 82,* 964 (1993).

[6] P.I. Lee. *J. Appl. Polym. Sci., 50,* 941 (1993).

[7] P. Vandewitte et al. *J. Controlled Release, 24,* 61 (1993).

[8] S. Narisawa, H. Yoshino, Y. Hirakawa, K. Noda. *Chem. Pharm. Bull., 42,* 1485 (1994).

[9] A.K. Banga, Y.W. Chien. *Pharm. Res., 10,* 697 (1993).

[10] R. Bowtell et al. *Magnetic Resonance Imaging, 12,* 361 (1994).

[11] S. Ueda et al. *J. Drug Targeting, 2,* 35 (1994).

[12] R. Rautenbach, R. Albrecht. *Membrane Processes.* John Wiley & Sons. New York (1989).

[13] K.O. Muranov, A.Ya. Polishchuk *et al. Khimiko-Farmatsevticheskii Zhurnal (Chem. Pharm. J.),* 24, 68 (1990) (In Russian).

[14] A.L. Iordanskii, A.Ya. Polishchuk, L.P. Razumovskii. *Indian J. Chem.,* 31A, 366 (1992).

[15] A.Ya. Polishchuk. *Intern. J. Polymeric Mater., 25,* 37 (1994).

[16] T.A. Zawodzinski. *J. Electrochem. Soc., 140,* 1981 (1993).

[17] G.M. Zentner, G.A. McClelland, S.C. Sutton. *J. Controlled Release, 16,* 237 (1991).

[18] M.D. Vyas, R.C. Mody, I.C. Mody. *J. Appl. Polym. Sci., 52,* 1031 (1994).

[19] B. Lindstedt, M. Sjoberg, J. Hjartstam. *Intern. J. Pharm., 67,* 21 (1991).

[20] G.E. Zaikov, V.S. Livshits. *Khimiko-Farmatsevticheskii Zhurnal (Chem. Pharm. J.), 18,* 586 (1984) (In Russian).

[21] V.S. Livshits, G.E. Zaikov. *Intern. J. Polymeric Mater., 16,* 277 (1992).

[22] N. Wakayama, K. Juni, M. Nakano. *Chem. Pharm. Bull., 29,* 3363 (1981).

[23] J. Heller. In: *CRC Critical Reviews in Therapeutics and Drug Systems,* J. Heller (Ed.), 1, p. 39 CRC Press, Boca Raton, Florida, (1984).

[24] X.C. Zhang et al. *J. Controlled Release, 31,* 129 (1994).

[25] K. Avgoustakis, J.R. Nixon. *Intern. J. Pharm., 99,* 247 (1993).

[26] S. Ueda et al. *J. Drug Targeting, 2,* 133 (1994).

[27] J.M. Marentette, A.E. Grosser. *J. Pharm. Sci., 81,* 318 (1992).

[28] G.D. Svoboda et al. *J. Controlled Release,* **20**, 195 (1992).
[29] G.D. Svoboda. *ACS Symp. Ser.,* **520**, 263 (1993).
[30] A.E. Grosser, M. Fitzsimons, L. Leonardi, J. Salha. *J. Pharm. Sci.,* **82**, 1061 (1993).
[31] J.R. Cardinal. In: *Medical Applications of Controlled Release,* R.S. Langer and D.L. Wise (Eds.), **1**, p. 41, CRC Press, Boca Raton, Florida (1984).
[32] P.I. Lee. *J. Controlled Release,* **4**, 1 (1986).
[33] A.Ya. Polishchuk et al. *Proc. 7-th All-Union Seminar on Protection of Handicrafts against Corrosion, Degradation and Biodegradation,* Minkhimprom, Moscow (1989) (In Russian).
[34] C.C. Chu. In: *CRC Critical Reviews in Biocompatibility,* **3**, p. 261, CRC Press, Boca Raton, Florida (1985).
[35] B.G. Amsden, Y.L. Cheng, M.F.A. Goosen. *J. Controlled Release,* **30**, 45 (1994).
[36] B.G. Amsden, Y.L. Cheng. *J. Controlled Release,* **31**, 21 (1994).
[37] M.R. Jenuqin et al. *J. Pharm. Sci.,* **81**, 983 (1992).
[38] Y. Kawashima et al. *Chem. Pharm. Bull.,* **40**, 196 (1992).
[39] S. Aoki et al. *Intern. J. Pharm.,* **85**, 29 (1992).
[40] R. Yoshida et al. *Polymer J.,* **23**, 1111 (1991).
[41] A. Gopferich, G. Lee. *Intern. J. Pharm.,* **71**, 237 (1991).
[42] E.M.G. Vanbommel et al. *Intern. J. Pharm.,* **72**, 19 (1991).
[43] N.F. Sheppard, M.Y. Madrid, R. Langer. *J. Appl. Polym. Sci.,* **46**, 19 (1992).
[44] M.P. Patel, M. Braden, S. Downes. *J. Mater. Sci.—Materials in Medicine,* **5**, 338 (1994).
[45] J.D. Smart. *Drug Develop. Industr. Pharm.,* **18**, 223 (1992).
[46] V.S. Livshits, A.V. Zakirov, V.A. Savin. *Proc. Symp. Polymers for Medicine,* p. 14, Zinantne, Riga (1975) (In Russian).
[47] Yu. V. Moiseev, G.E. Zaikov. *Chemical Stability of Polymers in Aggressive Media.* Khimiya, Moscow (1979) (In Russian).
[48] Yu. V. Moiseev, G.E. Zaikov. *Chemical Resistance of Polymers in Reactive Media.* Plenum Press, London (1984).
[49] N.M. Emanuel. *J. Polym. Sci. C.,* **67**, 195 (1980).
[50] G.E. Zaikov. *The Kinetics of Degradation of Some Polymers in the Living Body,* Report for the Commission on Medical and Biological Polymers of the USSR Academy of Sciences, Moscow (1979) (In Russian).
[51] G.E. Zaikov. In: *Proc. IV Intern. Conference on Advances in the Stabilization and Controlled Degradation of Polymers,* Lucerne, Switzerland, NY University Press (1982).
[52] G.E. Zaikov. In: *Proc. Conference on Degradation and Stabilization of Polymers,* p. 5, Tashkent, USSR Academy of Science Publ. (1976) (In Russian).
[53] G.E. Zaikov. In: *Proc. Conference on Catalysis in Organic and Bioorganic Chemistry,* p. 7, Frunze, USSR Academy of Science Publ. (1976) (In Russian).
[54] T.E. Rudakova, S.S. Kuleva, V.V. Pashkevichyus. *Polymer Science USSR,* **17**, 1550 (1974).
[55] L.V. Ivanova. *Thesis for the Degree of Candidate of Science (PhD Thesis),* Institute of Chemical Physics, USSR Academy of Sciences, Moscow (1974) (In Russian).
[56] Yu.V. Moiseev, V.S. Markin, G.E. Zaikov. *Plasticheskiye massy (Plastics),* **2**, 61 (1976) (In Russian).
[57] V.S. Livshits et al. *Vysolomolekulyarnye Soedineniya (Macromolecules),* **18B**, 336 (1976) (In Russian)
[58] K.Z. Gumargalieva, Yu. V. Moiseev and T.T. Daurova. *Principles of Biodegradation of Polymers,* Review of the USSR Ministry of Medical Industry, Moscow (1982) (In Russian).
[59] A. Joshi, K.J. Himmelstein. *J. Controlled Release,* **15**, 95 (1991).
[60] L. Disillvio et al. *Biomaterials,* **15**, 931 (1994).
[61] H. Okada et al. *J. Controlled Release,* **28**, 121 (1994).
[62] K.R. Kamath, K. Park. *Adv. Drug Deliv. Rev.,* **11**, 59 (1993).
[63] W.B. Dang, W.M. Saltzman. *J. Biomater. Sci. Polym. Ed.,* **6**, 297 (1994).

[64] J.F. Fitzgerald, O.I. Corrigan. *ACS Symp. Ser.*, **520**, 311 (1993).
[65] S.S. Shah, Y. Cha, C.G. Pitt. *J. Controlled Release*, **18**, 261 (1992).
[66] D. Peri et al. *Drug Develop. Industr. Pharm.*, **20**, 1341 (1994).
[67] P. Caliceti et al. *Farmaco*, **49**, 69 (1994).
[68] J.K. Lalla, K. Sapna. *J. Macroencapsulation*, **10**, 449 (1993).
[69] H.K. Sah, R. Toddywala, Y.W. Chien. *J. Controlled Release*, **30**, 201 (1994).
[70] A. Staubli et al. *JACS*, **112**, 4419 (1990).
[71] L.E. Cuebas et al. *J. Controlled Release*, **18**, 145 (1992).
[72] X. Xu, P.I. Lee. *Pharm. Res.*, **10**, 1144 (1993).
[73] Y. Aso, S. Yoshioka, A.L.W. Po, T. Terao. *J. Controlled Release*, **31**, 33 (1994).
[74] F. Fischel-Ghodsian, J.M. Newton. *J. Drug Targeting*, **1**, 51 (1993).

4 GENERAL APPROACHES TO REGULATION OF MULTICOMPONENT TRANSPORT IN POLYMER SYSTEMS FOR CONTROLLED RELEASE

The understanding of general regularities of multicomponent transport in polymers provides an opportunity to alter its kinetics and the rate. This chapter deals with the fundamental approaches to such an alteration in the systems of controlled delivery. In recent years new original methods were developed within the frame of the following approaches:

1. The synthesis of novel polymeric materials with regulated hydrophilic/hydrophobic balance.
2. The alteration of this balance by the controlled modification of polymers and/or use of fillers and additives.
3. The use of multilayer systems.

Of course, the methods of regulation of the behavior of the devices for controlled delivery are not limited by the above approaches. There is a huge number of other methods which serve many particular purposes. We should mention some of them such as Magnetism-Activated Drug Delivery, Ultrasound-Activated Drug Delivery, pH-Activated Drug Delivery, and Ion-Activated Drug Delivery which we do not consider here. The reader can find the details of their principles in the brilliant monograph edited by Lee and Good [1] and some other recent publications [2–7]. Also, we do not discuss the technological aspect of the design of devices although this problem is very important for engineers.

Therefore, our aim is not to review all possible methods of regulation of the controlled release, but to show how the theory of multicomponent transport can be applied for the development of some of these methods.

In the meantime, we could not pass by the biocompatibility of devices implanted in the body and would like to demonstrate that this very important problem can be also considered in terms of transport phenomena in polymers. This topic is discussed in the last paragraph of this chapter.

4.1. REGULATION OF POLYMER STRUCTURE

The alteration of hydrophilic/hydrophobic balance in copolymers by variation of the ratio between monomers or substitution of monomers is the obvious way to regulate the kinetics of water sorption and further release of the solute. Although the obvious does not mean the simplest, the possibility of such a regulation is the reason for the wide application of copolymers, polymer composites, and blends in the devices of controlled release.

We applied the theoretical approach which has been described in previous chapters for the explanation of the behavior of several copolymers. For example, films of vinylpyrrolidone/butyl methacrylate (VP/BM) and vinylpyrrolidone/methyl methacrylate (VP/MM) copolymers were investigated in [8]. Table 4.1 gives a list of their characteristics. The copolymers contained hydrophilic (N-vinylpyrrolidone) and hydrophobic groups that enable the thermodynamics and the kinetics of water sorption to be varied.

Table 4.1. Characteristics of copolymers of N-vinylpyrrolidone.

Property	Copolymers		
	VP/BM	VB/MM-1	VB/MM-2
Nitrogen content (%)	5.2–6.3	2.8–3.6	4.4
Relative viscosity	1.15–1.29	1.18–1.29	1.18–1.24
Relative strain (%)	250–300	80–100	4–6
Content of N-VP groups (g/g)	0.7	0.4	0.2
C_{w0} g/g	0.46	0.25	0.07
$D_{w0} \times 10^{-8}$ cm^2/s	7.5	4.2	2.0
$D_s(C_{w0}) \times 10^{-8}$ cm^2/s	50	6.2	0.4

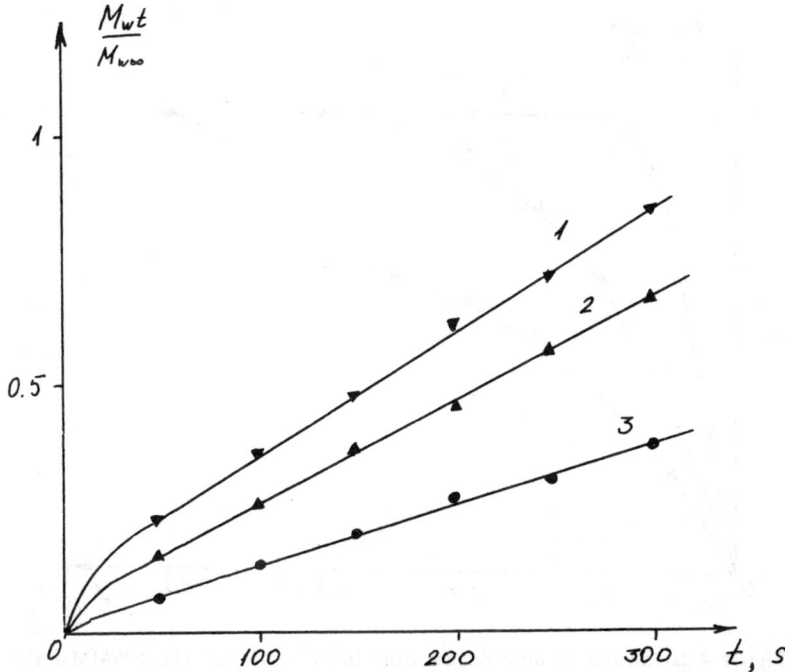

Figure 4.1. Kinetics of water sorption by copolymers of N-vinylpyrrolidone-VP/BM (1), VP/MM-1 (2), VP/MM-2 (3). $l = 100$ μ.

Table 4.1 and Fig. 4.1 show that equilibrium concentration of water and the rate of water sorption are the functions of the content of hydrophilic groups. The existence of the linear part of kinetics curves of water sorption may be explained in terms of the "generalized model of differential swelling stresses" which was reported in chapter 2 (Eqs. 2.3–2.11). The expected kinetics of the release of antiseptic drug, dioxidine, has been obtained for the films of all aforesaid N-VP copolymers. As pointed out in section 2.1.1 for high ratio between diffusion coefficients of solute $(D_s(C_{w0}))$ and water (D_{w0}) a drug release should follow water uptake. This feature is observed for VP/BM copolymer, and according to the kinetics of water sorption kinetic curve of dioxidine release out of this copolymer has a steady-state portion (Fig. 4.2). In agreement with the model the rate of release decreases as the above ratio reduces, showing non-Fickian diffusion for moderate $D_s(C_{w0})/D_{w0}$ value and the Fickian one for lower

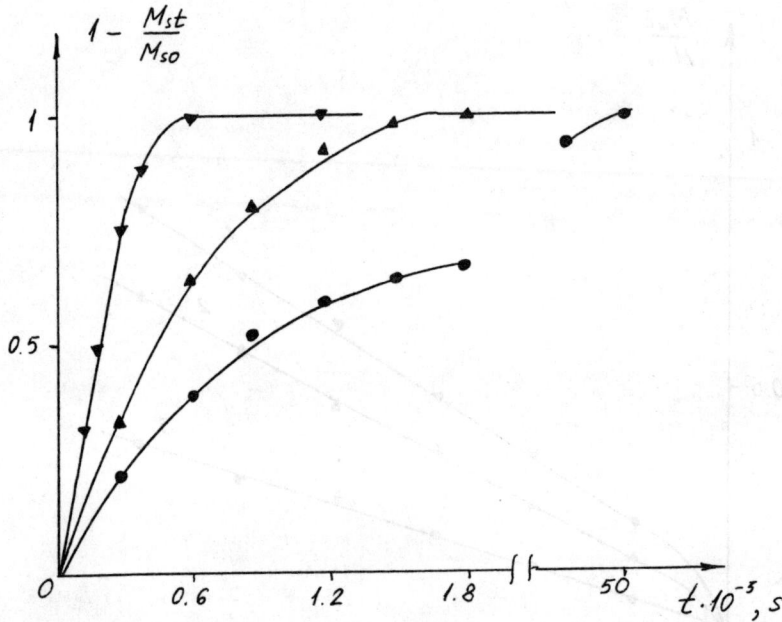

Figure 4.2. Kinetics of drug release from films of VP/BM (1), VP/MM-1 (2), VP/MM-2 (3) copolymers.

$D_s(C_{w0})/D_{w0}$ value which were obtained for VP/MM-1 and VP/MM-2 copolymers, respectively (Table 4.1).

Then, the model and experimental findings were applied for the preparation of fibers for surgery with programmed release of dioxidine. The good agreement between calculated kinetics of release from these fibers and those obtained in the clinical experiments is illustrated in Fig. 4.3.

A similar approach may be applied for a regulation of drug release from reservoir devices, the model of which is described in chapter 3 (see section 3.1.2). As Fig. 3.2 illustrates the steady-state release cannot be obtained for the case of negative exponential dependence of the diffusion coefficient of water on its concentration in the coating. We got this result experimentally (Fig. 4.4, curve 1) when we studied the release of salbutamol salt of benzoic acid from the reservoir coated by cellulose diacetate (CDA) [9]. The concentration dependence of the diffusion coefficient of water in CDA is quantitatively described by the equation

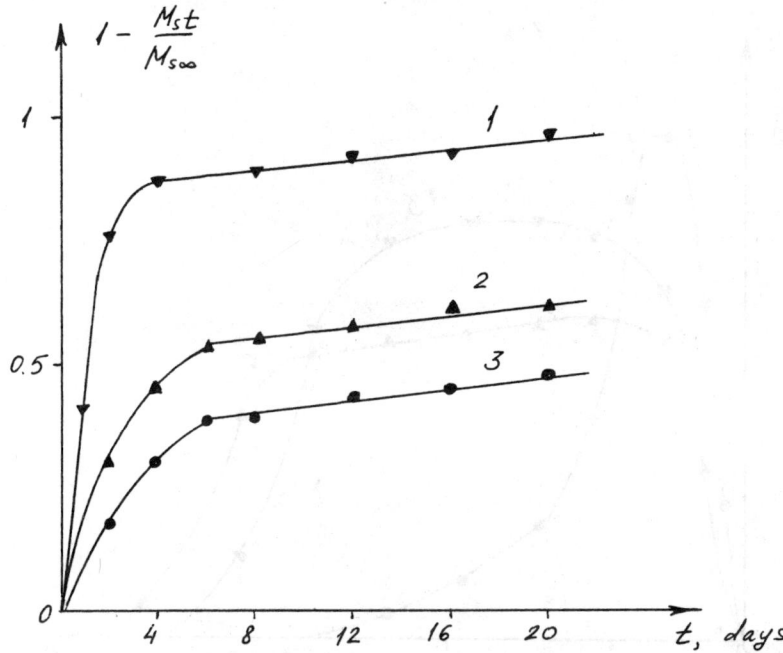

Figure 4.3. Kinetics of drug release from coatings of nylon fibers: VP/BM (1), VP/MM-1 (2), VP/MM-2 (3) fibers.

$$D_{w1} = 3.5 \times 10^{-8} \exp\left(-27C_w\right) \; [\text{cm}^2/\text{s}] \qquad (4.1)$$

where C_w is the concentration of water in the polymer of the dimension of [g/g]. Accordingly, the rate of salbutamol release passes through the sharp maximum, which is usually not desirable. Thus, we considered other cellulose derivatives to be more suitable for this device. It found that cellulose acetyl fluorate (CAF) and its blend with cellulose triacetate (CTA) are characterized by two opposite processes, which are swelling of polymer matrix causing the acceleration of transport of water and formation of clusters slowing it down. Isotherms of water sorption (Fig. 4.5) indicated slight plastification of these polymers. The analysis of these isotherms showed that the bimodal distribution of water in matrix should be expected in CAF and CAF + CTA with number of water molecules in clusters equal to 10 and 15, respectively. The main consequence of such a behavior of these

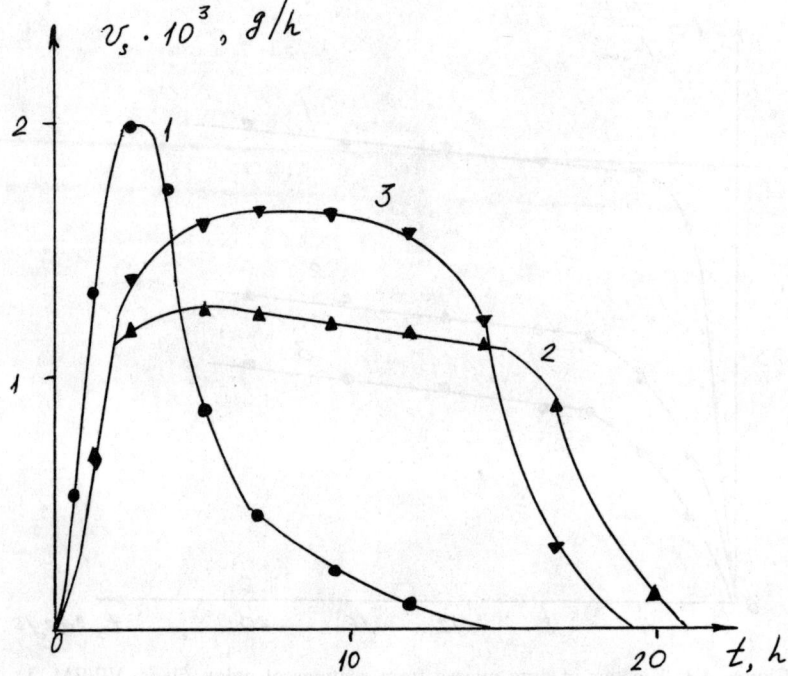

Figure 4.4. Kinetics of the release of salbutamol benzoate from the reservoir coated by CDA (1), CAF (2), and CAF+CTA (3).

polymer-water systems is the moderate increase of the diffusion coefficient of water which may be described by linear equation

$$D_{w2} = 8 \times 10^{-10} + 1.1 \times 10^{-7} \, C_w \; [cm^2/s] \qquad (4.2)$$

$$D_{w3} = 5.6 \times 10^{-9} + 3.3 \times 10^{-7} \, C_w \; [cm^2/s] \qquad (4.3)$$

for CAF and CAF + CTA, respectively.

Experimental results confirmed the model prediction which is described in section 3.1.2. The steady-state flux of salbutomol solution was reached for both coatings, and the time interval of this steady-state part was defined by constants involved in (4.2) and (4.3). Namely, the CAF + CTA coating provided steady-state release for longer time (Fig. 4.4, curve 3). The rate of release was higher for CAF (Fig. 4.4, curve 2) due to the higher value of water solubility in this polymer.

Polyurethane (PU) microspheres were prepared by a novel and simple

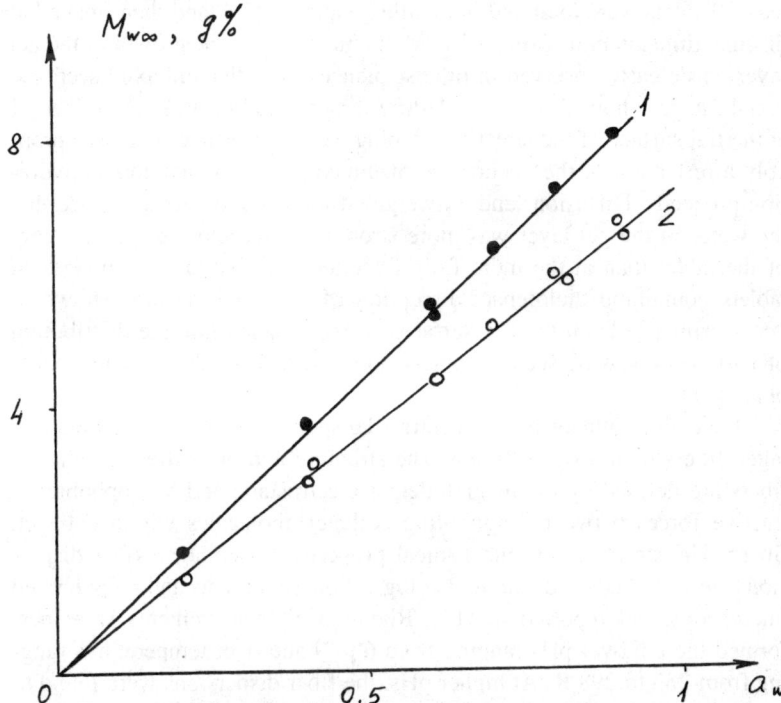

Figure 4.5. Water sorption isotherms for CAF (1) and CAF+CTA (2) films.

method using the condensation polymerization technique [10]. Microspheres of different morphological characteristics were prepared using toly-
lene 2,4-diisocyanate (TDI) and methylene diphenyl diisocyanate (MDI).
These microspheres were fully characterized by thermal analysis, scanning
electron microscopy, and spectral techniques. MDI-containing microspheres were found to be more porous as compared to TDI-containing microspheres. The infrared spectra indicated the complete utilization of the
isocyanate groups in the synthesis of microspheres. Bromothymol Blue
(BTB) was used as a model drug for the in vitro release studies from the
spheres. The results indicated that BTB released much faster using the MDI
spheres as compared to the TDI spheres.

NMR microscopy was used to monitor the formation of the gel layer
in hydrating hydrophilic polymer tablets. Such tablets were used in the controlled delivery of drugs, where it was found that the rate and extent of the
swelling of the outer gel layer critically influenced the kinetics of drug re-

lease. Tablets were hydrated in distilled water at 37°C and then imaged at discrete time intervals using a 500 MHz microscope. The growth of the gel layer was clearly observed in time sequences of radial and axial sections. Axial images showed some interesting dimensional changes, with the gel at the flat surface of the tablet developing a concave shape. This was probably a reflection of the occurrence of uniaxial stress relaxation as hydration proceeds. Diffusion- and T_2-weighted images provided evidence that the water in the gel layer was more strongly bound close to the dry core of the tablet than at the more fully hydrated outer surface. In images of tablets containing diclofenac, disruption of the gel layer was shown to occur primarily from the flat surfaces of the tablet, while the distribution of particles could be seen in tablets doped with insoluble calcium phosphate [11].

Injectable collagen is a concentrated dispersion of phase-separated collagen fibers in aqueous solution. The structure and properties of collagen fibers are defined by the magnitudes of electrostatic and hydrophobic attractive forces between neighboring collagen molecules within collagen fibers. The structure and mechanical properties of collagen fiber dispersions were studied by dynamic theological measurements and by polarized microscopy and reported in [12]. Rheological measurements were performed therein over pHs ranging from 6 to 9 and over temperatures ranging from 283 to 298 K. At higher pHs, the fiber dispersions were found to possess more rigid fibers and stronger inter-fiber attractive forces. This response is argued to result from changes in the ionization of amino acid side chains, which result in larger net-electrostatic attractive forces. Raising the temperature caused fibers to rigidify through enhanced hydrophobic attractive forces. Gels formed by lower pH–higher temperature fiber dispersions possess different properties than gels formed at higher pHs and lower temperatures.

Sah and Chien prepared a biodegradable microreservoir-type microcapsule for the controlled release of a model protein bovine serum albumin. BSA was incorporated into the microcapsules with high efficiency of 96.1%. The encapsulation did not cause any changes in the molecular weight and conformation of BSA, which was proven by biochemical analyses such as gel electrophoresis, circular dichroism, and HPLC. The compositions and fabrication technique of microcapsules were found to be closely related to the release of BSA from the microcapsules and their degradation. Depending on microcapsule formulations, the in vitro release profile of BSA was either monophasic or biphasic. The microcapsules provided various release rates. It was also possible to control the delay before the initiation of BSA release and total duration of its delivery [13].

Jou and Huang investigated the behavior of blends of rod-like PMDA-

B (pyromellitic dianhydride-benzidine) and semi-flexible 6FDA-PDA (6F-dianhydride-phenylendiamine) polyimides with several different compositions. According to the results from X-ray diffractometry and FTIR spectrometry, these two polyimides were incompatible when mixed at room temperature for 20 minutes. When mixed at 50°C for 40 hours, the polymers, thermodynamically incompatible, became compatible owing to exchange reactions. Bending beam diffusion experiments showed that the diffusion of moisture in these films belongs to case I. The diffusion coefficient D_w was 10^{-10} cm^2/s in PMDA-B and 1.65×10^{-9} cm^2/s in 6FDA-PDA. In the blends, D_w increased with increasing content of 6FDA-PDA. The slow diffusion in PMDA-B was attributed to its highly crystalline structure and relatively small interchain spacing. The diffusion of moisture was faster in the compatible films. Regardless of compatibility, diffusion in all the blends was much slower than in pure 6FDA-PDA. This was attributed to the comparatively small average interchain spacing of the blends [14].

The mass transport of two different compounds through polydimethylsiloxane (PDMS)-silica films was investigated by Dahl and Sue [15] for qualitative demonstration how this coating system can alter the release of various compounds. Various ratios of PDMS elastomer and silica were used to coat monodisperse particle-sized pellets layered with an ionizable compound (tartrazine) and a nonionized compound (acetaminophen). The 2:1 PDMS-silica composition containing the polyethylene glycol (PEG) 8000 pore former allowed mainly pore transport through void spaces in the PDMS films. Both compounds rapidly diffused through the film as a result of the solubilization and subsequent removal of the PEG 8000 from the film matrix. As the PDMS-silica ratios in the films changed from a 1:1 to a 2:1 to a 4:1 (all without PEG 8000) coating formulation, the differences in release rate between acetaminophen and tartrazine changed. The lower ratio of PDMS-silica allowed much faster tartrazine diffusion compared to acetaminophen. As the ratio increased from 1:1 to 2:1, the two compounds were released at similar rates. When the ratio reached 4:1, acetaminophen was released significantly faster than tartrazine. This study demonstrated that utilization of this polymer system offers a useful tool to modify release rates of ionic and nonionic drug substances.

Crystalline poly-D(-)-3-hydroxybutyrate (PHB) and its copolymers with poly-3-hydroxyvalerate [P(HB-HV)] are biodegradable polyhydroxyalkanoates with potential application in controlled drug delivery systems. Matrices of PHB and P(HB-HV) copolymers containing a model drug, methyl red, were prepared by solvent casting and melt-processing. Drug release from P(HB-HV) copolymer matrices produced by any given fabrication technique was dependent on copolymer composition. Progressively faster rates of drug release were obtained on increasing HV content. This could be explained by the different crystallization behavior of P(HB-HV) copolymers. Evidence from polarized light microscopy of copolymer spherulites suggested that copolymer films with increasing HV content exhibited morphologies with decreasing abilities to entrap the model drug [16].

The aggregation between hydrophobic/hydrophilic block copolymer unimers in water was analyzed by Petrak [17]. Based on the molecular weight and composition of the block polymer, the number of unimers in each particulate aggregate, the size of the particle and the size of the particle core, the average distance between the solvated chains attached to the micelle core was calculated. The distance between the terminally attached chains was compared with the Flory radius R(f) which was related to the extent that the chains were forced to stay in the "brush" conformation. By selecting the proper structure of block copolymer, micelles composed of water-soluble polymer chains were prepared with a "near ideal" saturated surface.

Various theoretical frameworks of analysis of drug transport in block and graft copolymers were examined by Harland and Peppas [18] and the importance of partially or totally impermeable domains was underlined therein. It was shown that the drug transport in and the release from block or graft copolymers containing hydrophilic and hydrophobic domains was governed by the size and shape of the domain and thermodynamic interactions between hydrophilic and hydrophobic components. This heterogeneous structure can be used to the advantage for the delivery of hydrophilic of hydrophobic drugs at desirable rates. On the basis of their theoretical approach, Harland and Peppas synthesized a series of hydrophilic/hydrophobic copolymers and characterized them for possible applications in drug delivery. Such systems were prepared by random, graft, or block copolymerization of various monomers, followed by selective hydrolysis. The copolymer composition was investigated using NMR spectroscopy, whereas their molecular weight distribution was determined by gel permeation chromatography. Thermodynamic and physical properties, such as transition temperatures and degrees of swelling were measured to elucidate the effect of phase separation on the formed network structure. Microdomain formations, balance of chemical properties, and network characteristics affected the partitioning and transport of theophylline and myoglobin in these swollen networks. Finally, the nonlinear network swelling, selective drug partitioning, and control of drug permeation through and release from copolymer membranes were correlated to the presence of hydrophobic and hydrophilic microdomains in the copolymer networks for a selected number of copolymers studied [19].

Copolymers of 2-hydroxyethyl methacrylate with a number of multiethylene glycol dimethacrylates were prepared by bulk copolymerization

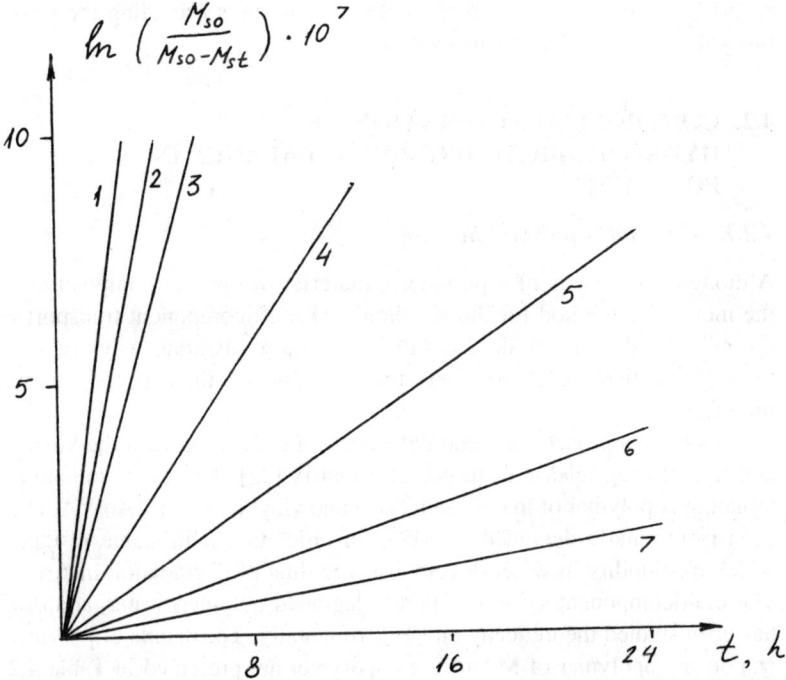

Figure 4.6. Effect of liquid crystal content on diffusion of salycilic acid at 80°C (1), 70°C (2), 60°C (3), 50°C (4), 40°C (5), 30°C (6), 20°C (7).

in the presence of AIBN as an initiator. The content of the dimethacrylate varied between 30 and 50% mol. Additional copolymers were prepared by solution polymerization in the presence of water or ethanol. All samples were swollen in water up to thermodynamic equilibrium, and their dynamic swelling behavior was studied as a function of time. It was concluded that the mechanism of water transport in these moderately and highly cross-linked polymers was a coupled relaxation and diffusion and that the relaxational contribution to the overall transport depended on the chain length of the multiethylene glycol dimethacrylate [20].

The permeability of solutes may be regulated by the formation of liquid crystal zones in polymer membrane [6]. As reported in [21] polysiloxanes with substituted side groups are able to form structures with a phase transition temperature in the interval of 45–65°C. Fig. 4.6 shows the effect of a mezogene phase of polysiloxanes on the rate of diffusion of salicylic acid in a wide temperature interval including that of phase transition. The increase of the content of a liquid crystal phase is shown to induce the reduction of the diffusion coefficient of solute, thus providing the possibility to govern the rate of its ejection.

4.2. CONTROLLED ALTERATION OF HYDROPHOBIC/HYDROPHILIC BALANCE IN POLYMER

4.2.1. Controlled polymer heating

Although a synthesis of a polymeric material of optimal composition is the most used method for the regulation of multicomponent transport in the controlled delivery devices, there are more accurate ways of controlled modification of polymer structure. One of them is the thermal modification.

Such an approach has been developed in order to govern the kinetics and rate of drug release from polymer matrix [22]. The water-soluble alternating copolymer of maleic anhydride and vinyl acetate (MAn/VAc) has been isothermally degraded at 205°C in order to obtain stable residues, which insolubility in water develops as the time of degradation increases. The multicomponent system of partly degraded polymer, water, and drug has been studied theoretically and experimentally. The details of polymerization of copolymer of MAn/VAc copolymer are presented in Table 4.2.

The thermal degradation of this copolymer has been studied by thermal volatilization analysis (TVA) [23]. Formation of conjugated double bonds and hydroxyl groups as well as the development of insolubility in water due to intermolecular dehydration are the most essential rearrangements that occurred under isothermal investigation at 205°C. Being related to the time of degradation, obtained data show that the first stage of degradation basically consists of acetic acid loss (Scheme 4.1) followed by subsequent elimination of carbon dioxide from the MAn (Scheme 4.2) units remaining in the chain.

Table 4.2. Conditions in the copolymerization of maleic anhydride with vinyl acetate at 60°C.

Total concentration of monomers	2.5 M
Total volume	85 ml
Initial mole fraction of MAn	0.5
Initiator (AIBN) concentration	10^{-2} M
Conversion	20%

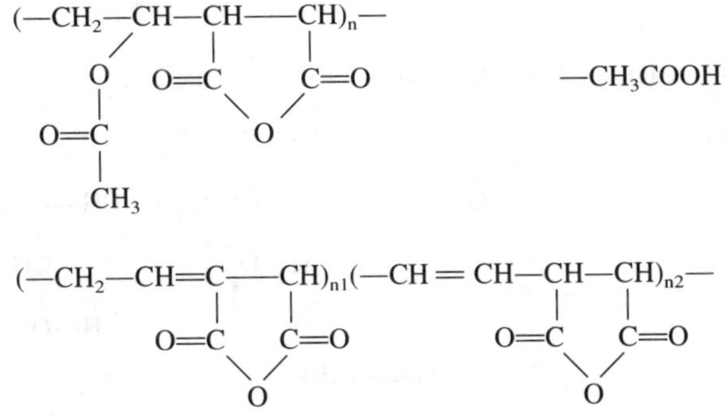

$$(-CH_2-CH-CH----CH)_n-$$

with pendant groups:
O / ; $O=C$; $C=O$ with O bridge; $O=C$; CH_3 ; $-CH_3COOH$

Scheme 4.1

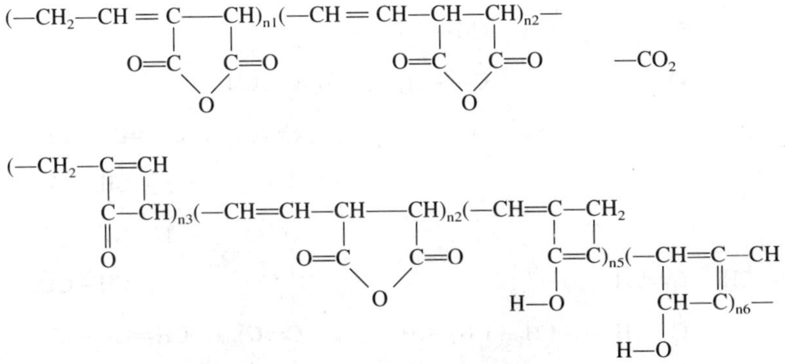

$$(-CH_2-CH=C----CH)_{n1}(-CH=CH-CH-CH)_{n2}-$$
$$O=C \quad C=O \qquad O=C \quad C=O$$
$$O \qquad\qquad O$$

$$(-CH_2-CH=C----CH)_{n1}(-CH=CH-CH-CH)_{n2}-$$
$$O=C \quad C=O \qquad O=C \quad C=O \qquad -CO_2$$
$$O \qquad\qquad O$$

$$(-CH_2-C=CH$$
$$C-CH)_{n3}(-CH=CH-CH----CH)_{n2}(-CH=C-CH_2$$
$$O \qquad\qquad O=C \quad C=O \qquad C=C)_{n5}(-CH=C-CH$$
$$O \qquad H-O \qquad CH-C)_{n6}-$$
$$H-O$$

Scheme 4.2

The CO_2 loss leads to the development of conjugation and finally to the total breakdown of the anhydride ring (Scheme 4.3).

The gradual insolubility develops partly due to the reduction in flexibility of the backbone, but also as a result of crosslinking by intermolecular dehydration involving OH groups produced in some of the structures (Scheme 4.4).

In chapter 1 (see section 1.1.2), we described the general conclusion regarding thermodynamics of water sorption by initial and partly degraded MAn/VAc copolymer. The sorption isotherms (Fig. 1.3) show how hydrophilic/hydrophobic balance in the matrix varies from water soluble (initial copolymer) to low-swelling polymer (40 min. of the degradation).

The concentration dependence of the diffusion coefficient of water in undegraded and partly degraded MAn/VAc copolymer (Table 4.3, Fig. 4.7)

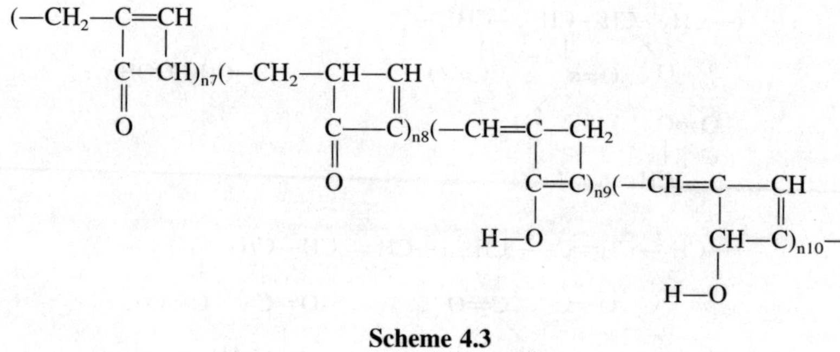

$$(-CH_2-C=CH$$
$$| \quad |$$
$$C-CH)_{n7}(-CH_2-CH-CH$$
$$|| \qquad\qquad\qquad ||$$
$$O \qquad\qquad C---C)_{n8}(-CH=C-CH_2$$
$$|| \qquad\qquad\qquad |$$
$$O \qquad\qquad\qquad C=C)_{n9}(-CH=C---CH$$
$$\qquad\qquad\qquad\qquad | \qquad\qquad\qquad\qquad ||$$
$$\qquad\qquad\qquad H-O \qquad\qquad\qquad CH-C)_{n10}-$$
$$\qquad\qquad\qquad\qquad\qquad\qquad\qquad\qquad |$$
$$\qquad\qquad\qquad\qquad\qquad\qquad\qquad\qquad H-O$$

Scheme 4.3

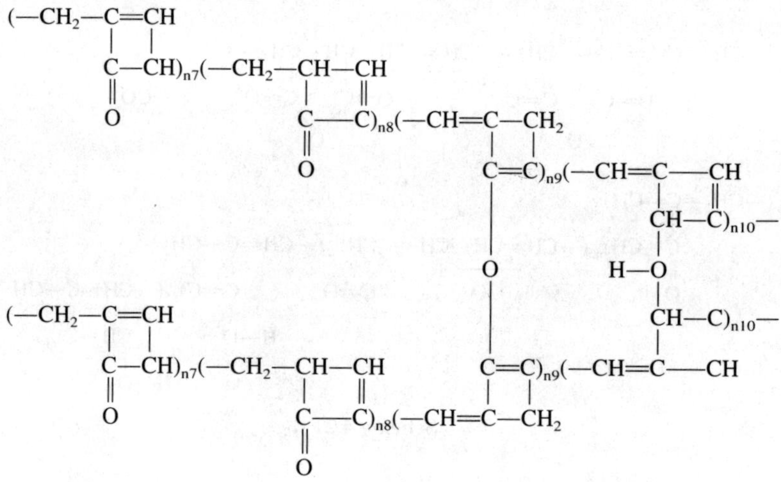

$$(-CH_2-C=CH$$
$$| \quad |$$
$$C-CH)_{n7}(-CH_2-CH-CH$$
$$|| \qquad\qquad\qquad ||$$
$$O \qquad\qquad C---C)_{n8}(-CH=C-CH_2$$
$$|| \qquad\qquad\qquad |$$
$$O \qquad\qquad\qquad C=C)_{n9}(-CH=C---CH$$
$$\qquad\qquad\qquad\qquad | \qquad\qquad\qquad\qquad CH-C)_{n10}-$$
$$\qquad\qquad\qquad\qquad O \qquad\qquad H-O$$

$$(-CH_2-C=CH$$
$$| \quad |$$
$$C-CH)_{n7}(-CH_2-CH-CH \qquad\qquad\qquad CH-C)_{n10}-$$
$$|| \qquad\qquad\qquad || \qquad C=C)_{n9}(-CH=C---CH$$
$$O \qquad\qquad C---C)_{n8}(-CH=C-CH_2$$
$$\qquad\qquad\qquad ||$$
$$\qquad\qquad\qquad O$$

Scheme 4.4

Table 4.3. Parameters concerned with the concentration dependence of the diffusion coefficient of water in initial and partly degraded MAn-VAc copolymer.

$$D_w = D_{wo}(f)\exp(k_{w1}C_w)$$

Parameter	Time of degradation (min.)						
	0	5	10	15	20	30	40
$D_{wo}10^8$ cm²/s	5.5	5.9	5.9	7.5	9.0	2.9	1.0
k_{w1}	5.3	1.8	0.3	0.12	0.02	−1.4	−2.1

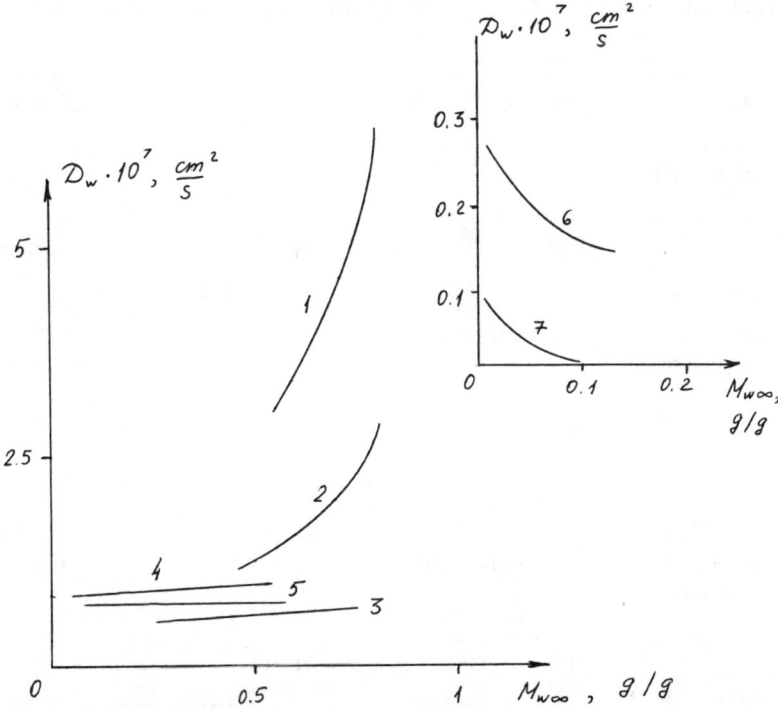

Figure 4.7. Dependence of the diffusion coefficient of water vapor on the concentration of water in MAn/VAc copolymer (1) and its residues after thermal degradation at 5 (2), 10 (3), 15 (4), 20 (5), 30 (6), 40 (7).

confirms different hydrophilic character of these materials. These dependencies put in the model (Eqs. 2.3–2.11) were used for the quantitative prediction of the regularities of controlled release of several drugs.

Among the parameters responsible for drug release from the system under study, the diffusion coefficient and solubility in water are the most important. The permeability of the partially degraded copolymer of MAn/VAc to the different drugs has been studied by the method described in [24]. Table 4.4 gives a list of these drugs and their solubility in water, and Table 4.5 illustrates the dependencies of the diffusion coefficients of these drugs on the time of degradation. The diffusion coefficient in the initial copolymer corresponds to the diffusion coefficient of the drug in pure water, and the dependence, as a whole, is the dependence of the diffusion coefficient of the drug on the concentration of water in the polymer, which should be put in the model (expression 2.11).

Table 4.4. Solubility of furan-containing drugs in water (concentration of the saturated solution).

Drug	Chemical Formula	$c_s^{\,o}$, mg/g
Furacilin, disassociate	O_2N furan ring $CH=N-N-C(=O)-NH_2$	5
Furagin, potassium salt	O_2N furan ring $CH=CH-CH=N-N-CH_2$, $O=C$ $C=O$, NK	2
Furacilin associated	O_2N furan ring $CH=N-N-C(=O)-NH_2$	0.25
Furadonin	O_2N furan ring $CH=N-N$ ring with CH_2, $C=O$, N, $C=O$	0.125

The effect of deceleration of drug release by controlled degradation of MAn/VAc copolymer has been predicted through calculations and proved by experimental data (Fig. 4.8).

The highly hydrophilic residues of MAn-VAc copolymer are characterized by high values of $D_s(C_{wo})/D_{wo}$. For these materials the expected ressult for drug release to follow water uptake was obtained. Neither solubility nor mobility of drug affect the kinetics of its release, except minor variations near to equilibrium.

The contribution of the solubility becomes visible for materials which contain from 20 to 50% of water in equilibrium, and the most remarkable for moderately swelling materials which contain no more than 10%. For the first group of the coatings (20–50%) the solubility of drug influences

Table 4.5. Dependence of the diffusion coefficient of drugs on time of thermal degradation and the corresponding value of overall water uptake.

		Time of degradation, min. (C_{wo}, g/g)						
		0 (oo)	5 (5.7)	10 (2.4)	15 (1)	20 (0.5)	30 (0.12)	40 (0.09)
		$D_s(C_{wo}) \times 10^7$ cm²/s						
	c^0_s, mg/g							
1	5	10	8.6	6.2	4.3	1.2	0.25	0.11
2	2	8.0	6.8	5.8	3.8	1.2	0.30	0.14
3	0.25	8.0	6.7	5.4	3.2	0.8	0.12	0.06
4	0.12	7.5	6.5	5.7	4.0	1.5	0.37	0.19

mostly the rate of release, while its kinetics remains essentially unchanged and includes the linear (case-II) part in the kinetic curve. We should emphasize, however, that this is not the effect of solubility only, but also of the mobility (relative diffusion coefficient) of the drug.

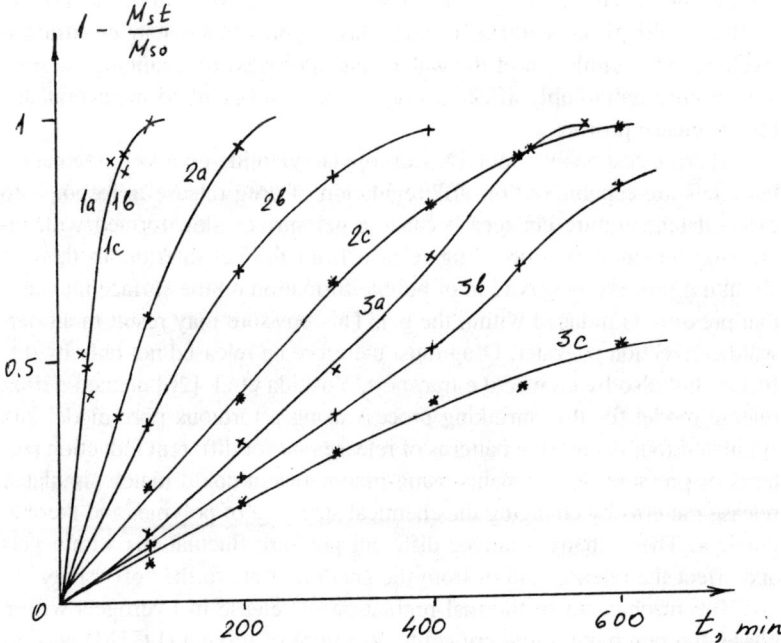

Figure 4.8. Computed and experimental kinetics of the release of furagin (a), furacilin associated (b), and furadonin (c) from partly degraded MAn/VAc copolymer after heating to 5 (1), 15 (2), and 30 (3) minutes

For low-swelling materials ($<$ 10%) the solubility affects both the rate and the kinetics of the release, and the latter is the effect of solubility only. The decrease of solubility finally leads to the Higuchi kinetics as the theory predicts.

Therefore, we get evidence that simultaneous control of chemical structure of polymer and its transport properties gives good opportunity to develop polymer systems for which required kinetics and rate of drug release can be obtained.

Thermo-responsive hydrogels of poly(N-isopropylacrylamide-co-butyl methacrylate) (poly-(IPAAm-co-BMA)) are capable of swelling–deswelling changes in response to external temperature. As poly(IPAAm-co-BMA) gels swell larger at a lower temperature, the degree and rate of the swelling could be controlled by temperature without altering the chemical structure. Therefore, drug release profiles were remarkably changed by alternation of temperature. The release profiles of indomethacin from poly(IPAAm-co-BMA) were observed to be zero-order at 20°C. Okuyama and coauthors explained this release [25] in terms of a case-II diffusion mechanism which indicates that relaxation of polymer chains with swelling was rate-determining. In the case of 10°C, release demonstrated a sigmodal profile. The acceleration of drug release was due to a rapid increase in swelling with disappearance of the glassy core which had constrained swelling. The regulation of the water-uptake process by changing external temperature remarkably affected drug release and resulted in several different release profiles.

Thermo-responsive poly(N-isopropylacrylamide-co-alkyl methacrylate) gels are capable of "on-off" regulation of drug release in response to external temperature changes because a gel surface skin formed with increasing temperature stops drug release from the gel interior. In this gel shrinking process, observation of bubble formation on the surface indicates that pressure is induced within the gel. This pressure may result in an outward convection of water. Drug must therefore be released not only by diffusion, but also by convective transport. Yoshida et al. [26] created a drug release model for this shrinking process using a tortuous pore model and simulated four decreasing patterns of release rate for different induction patterns of pressure. Experiments using indomethacin could match simulated release patterns by changing the chemical structure of polymer and thermal gradient. These changes induce different pressure fluctuations within gels and affect the release pattern from the gel "on" state to the "off" state.

The mechanism of thermal regulation of release in hydrogels with a phase diagram having low critical temperature of mixing (LCTM) was reported in [6]. When the thermally reversible gel passes LCTM under heating or cooling the polymer matrix is contracted or expanded. This results

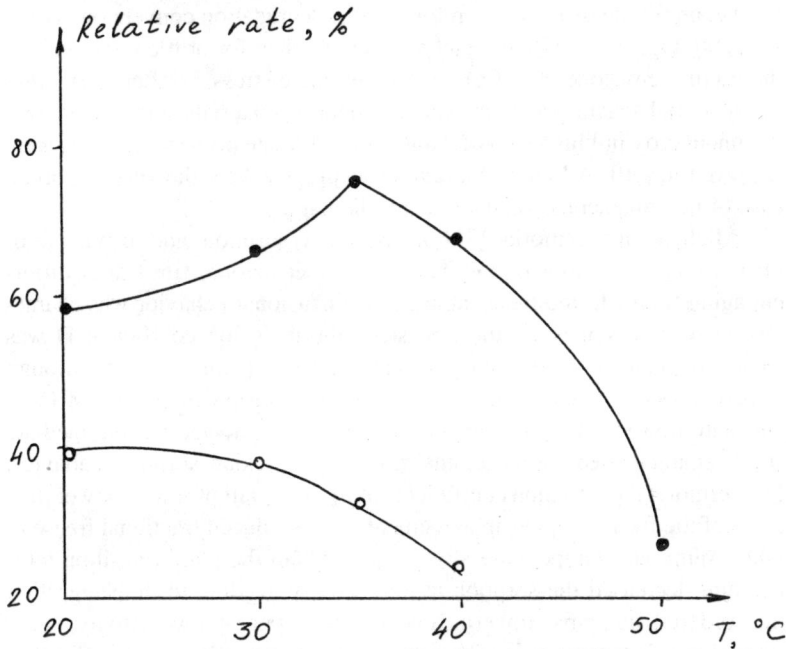

Figure 4.9. Relative rate of the release of methylene blue from poly-N-isopropylacrylamide gel in citric phosphate buffer (pH = 7.4) (1) or in deionized water (2).

in the desorption of the previously sorbed compounds [27, 28] or their extra sorption. This mechanism works when toxins are being ejected from physiological medium [28] or enzymes are being immobilized in a gel [29]. Fig. 4.9 illustrates this effect for release of the model compound (methylene blue) from hydrogels of 1:1 poly(N-isopropylacrylamide-co-methacrylic acid) with LCTM near 40°C [28].

Aging can modify polymer structure at the molecular, macromolecular, and/or the morphological level and thus induce changes in the mechanical properties. Stiffness is generally not modified for nonrubbery materials, except for mass transfer (solvent plasticization or plasticizer loss) in amorphous polymers or phase transfer (crystallization or crystal destruction) in semicrystalline polymers. The most significant modulus changes occur in the radiochemical aging of semicrystalline polymers whose amorphous phase is in the rubbery state. Yield properties generally vary in the same way as stiffness. Physical aging at $T < T_g$ can lead to a significant increase in the yield stress. Very general features can be observed for rupture properties, for instance: (1) Only ultimate elongation E is a pertinent variable in kinetic studies of aging involving tensile testing and related methods,

(2) the amplitude of E variation for a given degradation conversion is considerably higher for initially ductile materials than for brittle ones, and (3) the rupture envelope $\sigma = f(\varepsilon)$, i.e., the ultimate stress, is often very close to the initial tensile curve except for rubbery materials undergoing predominant crosslinking. The mechanisms of ultimate property changes were reviewed in [30]. A kinetic approach was proposed for the very important case of heterogeneous, diffusion-controlled aging.

Michele and Vittoria [31] measured a sorption and diffusion of dichloromethane vapor in poly(aryl ether ether ketone) films after different aging times. In the fresh samples the diffusional behavior was characterized by two stages: In the first stage the diffusion coefficient D was weakly dependent on concentration, and transport mainly occurred through a frozen system; whereas in the second stage a steep dependence of D on concentration was found. The aged samples also showed an intermediate stage, characterized by an anomalous or non-Fickian sorption behavior. Furthermore, the diffusion coefficients of the aged samples were lower than those of the fresh samples, in agreement with a reduced fractional free volume. Aging at a temperature slightly lower than the glass transition temperature decreased the sorption at low activity, leading to the suggestion that ordered domains, impermeable to the vapor at low activity, were formed at this temperature. The process of solvent-induced crystallization on the amorphous samples was investigated, and it was found that crystallization is induced after activity 0.7. Both the fresh and the aged samples showed the same behavior, although the level of crystallinity attained was found to be higher for the latter samples [31].

The hydraulic permeation of water through heat-treated polyvinyl alcohol membranes was investigated by Yang and Chu [32]. They found that the transport of water in a heat-treated polyvinyl alcohol membrane follows the mechanism of solution diffusion equation. Their results showed that the activation energy of water diffusivity increased and water diffusivity decreased with increase in the period of heat treatment. The compressibility of polyvinyl alcohol membranes decreased with increase in the period of heat treatment.

4.2.2. Controlled load of solute, fillers, and additives

Although there is a principle possibility to govern the water sorption in and solute release from the polymer matrix by the load of fillers and additives and some theoretical and experimental work has been done in this area (see 2.2.3), the application of this method has received so far little discussion. This is, most probably, due to the toxic character of many chemicals usually used as additives. For this reason their possible release together with the main components is not desirable and, mainly, water soluble ingredients are used to regulate the kinetics and the rate of release.

The effect of additives on the release of drugs from polyacrylic emulsion has been studied for its possible application as skin bandages. We studied the simultaneous release of drugs of different solubility and the effect of a water soluble ingredient, sodium alginate, on the drug transport. For this particular system the effect of additives (including drugs) on the equilibrium concentration of water in the polymer is described by the equation

$$C_{w0} = C^0_{wo}\exp(K_{w1}C_{s1} + K_{w2}C_{s2} + K_{w3}C_{s3})$$

where C^0_{wo} is the equilibrium concentration of water in unfilled polymer ($C^0_{wo} = 0.5$ g/g); C_{s1}, C_{s2}, and C_{s3} are the concentration of dioxidine, furacilin, and sodium alginate in the polymer, respectively; and K_{w1}, K_{w2}, and K_{w3} are the thermodynamic constants equal to 1, 3.6, and 12 l/mol, respectively. In the meantime, the kinetics of water sorption and its rate remain unchanged ($D_w = 2.3\times10^{-7}$ cm²/s) for variations of the drug load in a wide range, and minor changes are observed if sodium alginate is incorporated in the emulsion.

Since the kinetics and rate of solute release are affected by water concentration in the polymer, the programmed regulation of release is possible as it is shown in Fig. 4.10a. The main regularities observed are the following. In the absence of sodium alginate the diffusion of dioxidine may be described in terms of Fickian diffusion

$$\delta C_{s1}/\delta t = (\delta/\delta x)(D_{s1}\delta C_{s1}/\delta x)$$

where

$$D_{s1} = 9.2 \times 10^{-6} \exp(-2.4/C_{wi}) \text{ [cm}^2\text{/s];}$$

$$D_{s1}(C_{wo}) = 6.3 \times 10^{-8} \text{ [cm}^2\text{/s].}$$

The release of the drug of limited solubility (furacilin) follows the Higuchi kinetics with effective diffusion coefficient

$$D_{s1} = 5\times10^{-11} \exp(-2.4/C_{wi}) \text{ [cm}^2\text{/s].}$$

The load of the water soluble filler (sodium alginate) accelerates the release of both drugs and alters the kinetics of the desorption of furacilin in agreement with the theory described in chapter 2 (Fig. 4.10b). The dependence of diffusion coefficient of dioxidine changes in the presence of sodium alginate to

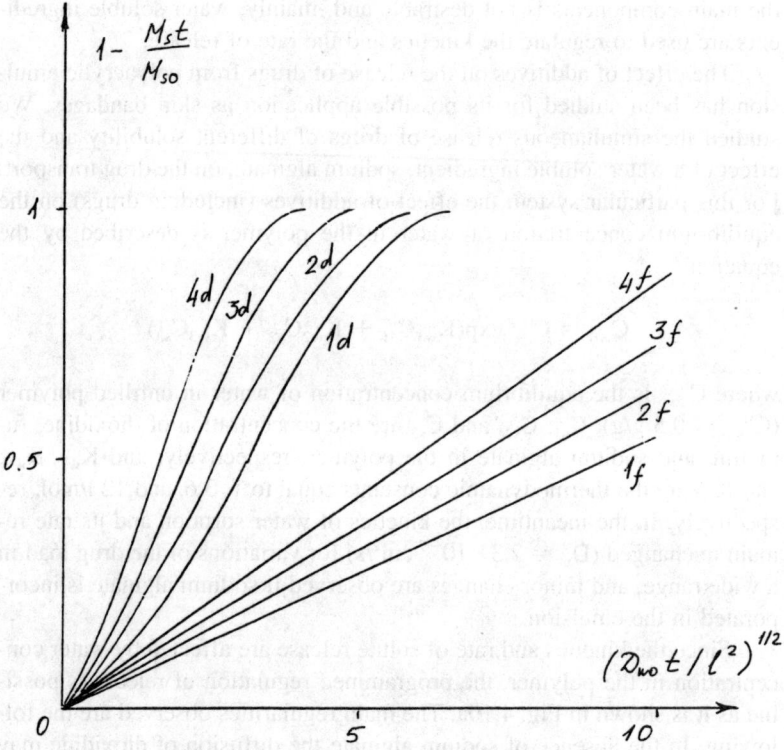

Figure 4.10a. Kinetics of release of dioxidine (d) and furacilin (f) from polyacrilic emulsion at different initial ratio between drugs loaded: $C_{s1o}/C_{s2o} = 10{:}2$ (1), 5:5 (2), 2:4 (3); 1:5 (4). $C_{s3o} = 0$ g/g.

$$D_{s1} = 9.2 \times 10^{-6} \exp(-2.4/C_{wi}) \; [\text{cm}^2/\text{s}];$$

$$D_{s1}(C_{wo}) = 5 \times 10^{-7} \; [\text{cm}^2/\text{s}].$$

Therefore, we get evidence that the load of a water soluble ingredient is one approach to the regulation of the kinetics of release of incorporated solutes.

Schroen et al. [33] studied problems encountered with an emulsion/membrane bioreactor. In this reactor, enzyme- (lipase) catalyzed hydrolysis in an emulsion was combined with two in-line separation steps. One was carried out with a hydrophilic membrane, to separate the water phase, the other with a hydrophobic membrane, to separate the oil phase. In the absence of enzyme, sunflower oil/water emulsions with an oil fraction between 0.3 and 0.7 could be separated with both membranes operating si-

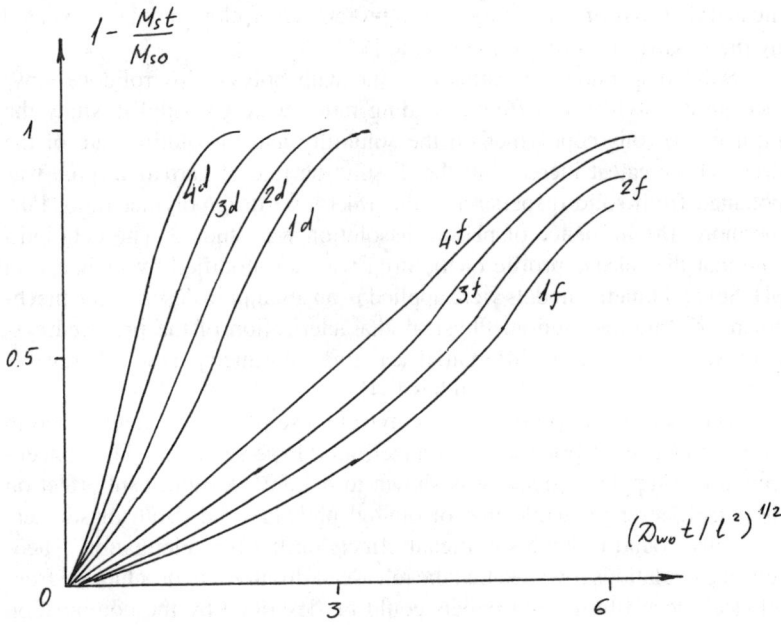

Figure 4.10b. Kinetics of release of dioxidine (d) and furacilin (f) from polyacrilic emulsion at different initial ratio between drugs loaded: C_{s10}/C_{s20} = 10:2 (1), 5:5 (2), 2:4 (3); 1:5 (4). C_{s30} = 0.04 g/g.

multaneously. However, two problems arose with emulsions containing lipase. First, the flux through both the hydrophilic and the hydrophobic membranes decreased with exposure to the enzyme. Second, the hydrophobic membrane showed a loss of selectivity demonstrated by permeation of both the oil phase and the water phase through the hydrophobic membrane at low transmembrane pressure. These phenomena were explained by protein (i.e., lipase) adsorption to the polymer surface within the pores of the membrane. It was proven that lipase was present at the hydrophilic membrane and that this, in part, explains the flux decrease of the hydrophilic membrane. To prevent the observed loss of selectivity with exposure to protein, the hydrophobic polypropylene membrane (Enka) was modified with block copolymers of propylene oxide (PO) and ethylene oxide (EO). These block copolymers act as a steric hindrance for proteins that come near the surface. The modification was successful: After 10 days of continuous operation the minimum transmembrane pressure at which water could permeate through an F 108-modified membrane was 0.5 bar, the same value as that observed in the beginning of the experiment. This indicated that loss of selectivity due to protein adsorption may be prevented by the modification of the membrane [33].

Solid dispersions of carbamazepine with polyvinylpyrrolidone/vinylacetate copolymer at different loading ratios were prepared to study the influence of this copolymer on the solubility and dissolution rate of the drug. The greatest increase in the dissolution rate of carbamazepine was obtained from solid dispersion at the 1/4 (w/w) drug/polymer ratio. Furthermore, the influence of pH on dissolution was studied. The data indicate that the release profile of the drug was not modified by a change in pH. Several kinetic models were applied in an attempt to describe the mechanism of drug dissolution. Physical characterization of the prepared systems was carried out by differential scanning calorimetry (DSC), X-ray diffractometry, and wettability studies [34].

The mechanical properties of ethylcellulose films were determined to evaluate effects of polymer, plasticizer, and dispersed solid. The concentration of propylene glycol was shown to exhibit no significant effect on the tensile strength while that of diethyl phthalate did. Both plasticizers were also found to have significant effects on the film elongation. Theoretical calculations revealed that the release of diltiazem hydrochloride from ethylcellulose film-coated pellets could be described by the combination of constant and nonconstant activity source diffusion-controlled model. A dramatic modification in drug-release characteristics was observed after the film-coated pellets had been compressed into tablets. However, the increase in compression force within the working range (181.8, 272.7, and 363.6 kg) was distinctly found to slightly affect the release from pellets being compressed. These behaviors may not be caused by flaws or failures within the film, but instead by the appreciable alterations in some physical properties of the film itself under pressure. The electron photomicrographs confirmed such a hypothesis [35].

Adhesive dispersion-type transdermal drug delivery (a-TDD) systems consisting of a monolayer of drug-loaded adhesive matrix were developed from three types of silicone-based pressure-sensitive adhesives. The adhesive polymers were tailored such that two of them were lipophilic (Bio-PSA® X7-2920 and Dow Corning® -355 Medical Adhesive) and one was relatively hydrophilic (E8086® adhesive) in nature. Three steroids, namely, progesterone, testosterone, and hydrocortisone, were used as model penetrants, and their release from the a-TDD systems and permeation through skin were investigated. The adhesive properties of these systems were studied, and the partial and total solubility parameters of these adhesive polymers were determined. The release of steroid molecules was observed to be a complex function of the physicochemical properties of the drug and polymer. The adhesiveness as determined from a standard peel test indicated that incorporation of the drug in higher drug loading doses results in a loss of adhesiveness. The results suggest that the chemical nature of the polymer is an important consideration when studying such adhesives for transdermal drug delivery [36].

4.2.3. Multilayer systems

Many materials which are widely used in medicine, agriculture, and industry are not suitable enough when their application for controlled delivery of chemicals (mainly drugs) is required. On the other hand, there are a dozen polymers with appropriate transport characteristics which, however, cannot be applied themselves as fibers, bandages, etc., due to the lack of their mechanical and other properties. In these cases, the development of multilayer systems allows us to solve the above problem. In the meantime, the application of multilayer systems is not only a technological problem. It is one of the ways to regulate the kinetics and the rate of release.

We particularly studied the multilayer system which included nylon coated by blend of MAn/VAc copolymer with drug [22]. The hydrophilic/hydrophobic balance in the second (copolymer) layer was varied by controlled isothermal heating as described in section 4.2.1. Redistribution of drug between two polymers defines multistage process of its further release, which includes desorption from moderately hydrophilic polyamide being in contact with solvent followed by the transport in swelling or dissolving MAn-VAc copolymer.

Taking into account these stages, we used the following equations to describe multicomponent transport in each layer

$$\delta C_{wi}/\delta t = (\delta/\delta x)(D_{wi}\delta C_{wi}/\delta x) \qquad 0 < x < 1 \qquad (4.4)$$

with

$$D_{wi} = D_{woi}\exp(k_{w1i}C_w + k_{w2i}f) \qquad (4.5)$$

and

$$\delta C_{si}/\delta t = (\delta/\delta x)(D_{si}S_{si}\delta a_{si}/\delta x) \qquad 0 < x < 1 \qquad (4.6)$$

with

$$D_{si} = D_{soi}\exp(-k_{w3i}/(C_{wi}+k_{w4i})) \qquad (4.7)$$

$$S_{si} = C_{si}/a_{si} = k_{ssi}C_{wi}V_w{}_ic_s{}^\circ = K_{si}C_{wi} \qquad (4.8)$$

where the main definitions are the same as in chapter 2, where the generalized swelling stress model has been reported, and index i corresponds to the number of layer (1 or 2).

The boundary and initial conditions for this multilayer system were

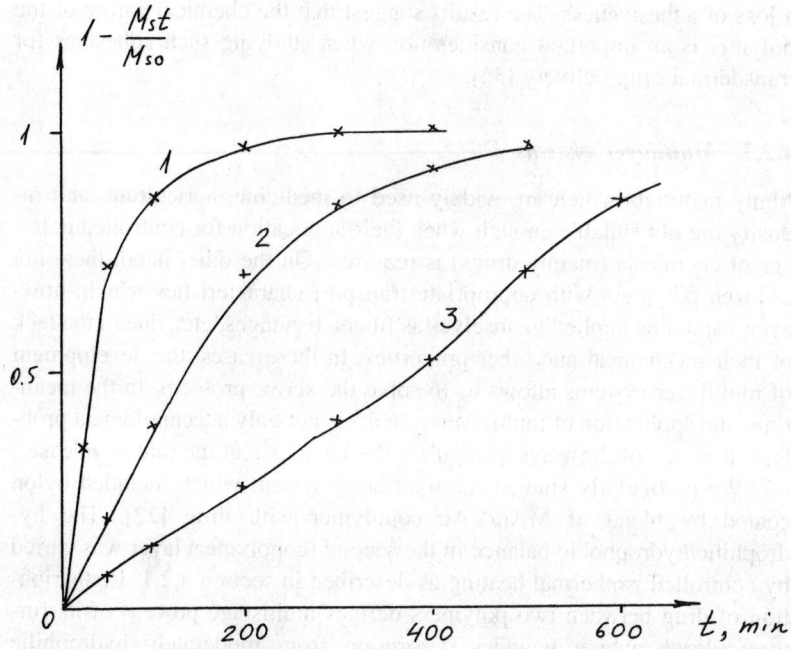

Figure 4.11. Kinetics of furacilin release from a polyamide textile bandage through films of partly degraded MAn/VAc copolymer after heating to 5 (1), 15 (2), and 30 (3) minutes.

selected as follows (x=0 corresponds to the nylon/water interface, x=1 corresponds to the nylon/MAn-VAc interface, and x=L corresponds to MAn-VAc water interface).

$$D_{w1}(\delta C_{w1}(l,t)/\delta x) = D_{w2}(\delta C_{w2}(l,t)/\delta x), \quad C_{w1}(l,t) = \alpha_w C_{w2}(l,t) \quad (4.10)$$

$$D_{s1}(\delta C_{s1}(l,t)/\delta x) = D_{s2}(\delta C_{s2}(l,t)/\delta x), \quad C_{s1}(l,t) = \alpha_s C_{s2}(l,t) \quad (4.11)$$

$$C_{w1}(0,t) = C_{w1o}; \qquad\qquad C_{w2}(L,t) = C_{w2o} \qquad (4.12)$$

$$C_w(x,0) = 0; \qquad\qquad 0 < x < L \qquad (4.13)$$

$$C_s(x,0) = 0 \qquad\qquad 0 < x < l \qquad (4.14)$$

$$C_s(x,0) = C_{so} \qquad\qquad l < x < L \qquad (4.15)$$

where α_w and α_s are the distribution coefficients between two layers for water and solute, respectively.

The effect of the first layer on the kinetics of release of a model drug is visible near to equilibrium, mainly, for highly swelling coating as illustrated in Fig. 4.11 in comparison with Fig. 4.8. These variations become minor and disappear as the hydrophobicity of the MAn/VAc coating (due to the programmed degradation) develops.

The aforesaid example relates to the systems when the second layer initially contains the solute. These systems are mostly applied in transdermal devices or other kinds of skin bandages [24, 37].

If the solute is initially loaded in the first layer, we must substitute the initial conditions (4.13–4.15) by

$$C_w(x,0) = 0; \qquad\qquad 0 < x < L \qquad\qquad (4.16)$$

$$C_s(x,0) = C_{s0} \qquad\qquad 0 < x < 1 \qquad\qquad (4.17)$$

$$C_s(x,0) = 0 \qquad\qquad 1 < x < L \qquad\qquad (4.18)$$

Composition and thickness of the second layer were varied in order to alter the kinetics of the release of dioxidine from the above-mentioned VP/MM films. The agreement between theoretical and experimental data showed (Fig. 4.12) that the transport of water and solute is controlled by the diffusion in the layers and is not substantially affected by the interlayer

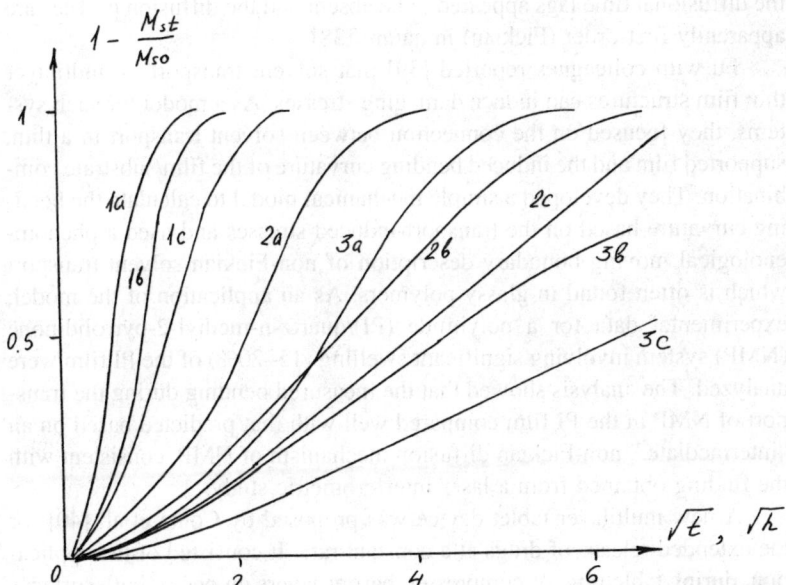

Figure 4.12. Kinetics of dioxidine release from the films of VP/MM-1 films coated by the second layer of VP/BM (1), VP/MM-2 (2), and polyethoxicyanacrylate (3). Thickness of the second layer = 50 (a), 80 (b), 120 (c) μ.

processes. The fundamental results were then applied for the development of VP/MM coatings for fibers with the programmed release of dioxidine.

A new method of preparing composite poly(vinyl alcohol) (PVA) beads with a double-layer structure was developed by Kim and Lee [38] and involved a stepwise saponification of suspension polymerized poly(vinyl acetate) (PVAc) beads and subsequent stepwise cross-linking of the PVA core and shell with glutaraldehyde. This process results in PVA beads with thin, highly cross-linked outer shells and lightly cross-linked inner cores of different degrees of cross-linking. In addition to structural characterization of the polymer based on equilibrium swelling measurements, the kinetics of water swelling and drug release from these beads were studied at 37°C using acetaminophen and proxyphylline as model drugs. The results showed that the outer shell functions as a rate-controlling membrane upon increasing its cross-linking ratio, X, above 0.47. This aspect is reflected in the observed diffusional time lags and constant-rate regions during swelling and drug release. Based on observed time lags, the diffusion coefficient of water through the outer PVA shell with a high cross-linking ratio of X = 0.5 was estimated to be at least six times higher than that of acetaminophen and proxyphylline. In addition, drug diffusion coefficients in the lightly cross-linked PVA core appeared to be at least 10 times larger than that in the highly cross-linked outer shell. At lower shell cross-linking ratios (X < 0.4), the diffusional time lags appeared to be absent and the diffusion profiles are apparently first-order (Fickian) in nature [38].

Fu with colleagues reported [39] that solvent transport in multilayer thin film structures can induce damaging stresses. As a model for such systems, they focused on the connection between solvent transport in a thin, supported film and the induced bending curvature of the film/substrate combination. They developed a simple mechanical model to calculate the bending curvature based on the transport-induced stresses and used a phenomenological moving boundary description of non-Fickian solvent transport which is often found in glassy polymers. As an application of the model, experimental data for a polyimide (PI)/quartz-n-methyl-2-pyrrolidinone (NMP) system involving significant swelling (15–20%) of the PI film were analyzed. The analysis showed that the measured bending during the transport of NMP in the PI film compared well with that predicted based on an "intermediate," non-Fickian diffusion mechanism of NMP, consistent with the finding obtained from a laser interferometric study.

A new multilayer tablet device was proposed by Conte et al. [40] for the extended release of drugs at a constant rate. It consisted of the application, during tabletting, of compressed barrier layers on one or both surfaces of a hydrophilic matrix containing the active ingredient. The barriers can limit the core hydration process by maintaining the planar surfaces of the tablet covered during dissolution. Moreover, the coatings reduce the extent of the core surface exposed to the interaction with the dissolution medium and to drug release. As a result, the release rate was reduced and the kinetics shifted towards constant drug release. The effects of the coating application on the release patterns of three different matrix cores containing trapidil or sodium diclofenac as model drugs were tested. The matrix cores were designed to provide different release mechanisms and extents.

Latex film matrix systems with a nonuniform drug distribution were prepared by Li and Tu for a coating process. In this process a drug concentration gradient in the coating dispersion was generated by the programmable pumping of the latex into a drug reservoir which contained the drug and latex. The film matrix formed by the dispersion would have a built-in drug concentration gradient as the coating process proceeded. A mathematical model was developed to describe the concentration change of the active ingredient in the coating dispersion as a function of the spraying rate of the coating dispersion, the pumping rate of the latex into the drug reservoir, and the initial drug concentration in the dispersion. The applicability of this process was demonstrated by the controlled in vitro dissolution of acetaminophen from ethylcellulose latex film matrices formed on glass beads. The release profile of the active ingredient from the systems changed as the drug concentration gradient profile in the matrix was altered, and a higher drug concentration gradient in the matrix resulted in a slower release rate and a more linear release profile [41].

Wan and Lai developed a new model system for controlled release using multilayer coated granules prepared by consecutively spraying aqueous solutions of diphenhydramine and HCl, dissolved in methylcellulose (MC) of various viscosity grades, onto lactose granules in a fluidized bed. The in vitro drug release profiles of the coated product were shown to be a function of the polymer viscosity grade and the sequential order of application of the polymeric component layers on the lactose granules. The permutation and the mass fraction of the component layers on the multilayer coated granules was altered to provide different drug release rates. Although the equilibrium hydration and swelling of each component layer influenced drug diffusion to some extent, the overall release was not entirely governed by it. The drug-release profiles were also modulated by the dissolution characteristics, particularly the particle size of the recrystallized drug that is embedded in the polymeric coat. This approach on controlled-release provides an outline for further work in this field [42].

Physical mixtures of an acrylate methacrylate copolymer (in fine powder) and salicyclic acid (drug model) in a ratio of 1:4 were compressed directly (i.e., without further granulation) to matrix (nondisintegrating) tablets to serve as core models. A special technique was used to form an aqueous-based coating system consisting of a water-insoluble copolymer and sucrose in varying proportions. Film coatings were applied on the matrix cores to a thickness of 75 μ (SEM). Drug-release rates increased as a linear function of the initial sucrose content in the film coatings. The amount of sucrose leached from the film coatings into the leaching fluid also increased linearly and in parallel with the increase in release rates. The release profiles obtained with cores with coating thickness of 75 μ were generally without a time lag. Doubling the coating thickness introduced a time lag (preceding the onset of release) of about 2.5 hours, 1.3 hours, or 0.5 hour for polymeric coatings without sucrose, or with sucrose 12.5% w/w, or 25% w/w, respectively. Coatings with a higher sucrose content did not display the release time lag. Overall release rates were hardly affected by the increase in coating thickness [43].

Yuen, Deshmukh, and Newton recently gave a good example of a combined application of many approaches described above. They developed a multiparticulate sustained release formulation of theophylline and evaluated it in vitro. The formulation comprised spherical pellets of high drug loading, coated with a rate-controlling membrane. The pellets were prepared using an extrusion spheronization method, while coating was performed with an aqueous dispersion of ethylcellulose using a fluidized bed coating technique. When ethylcellulose was used alone as the coating polymer, the drug-release profile was unsatisfactory, but could be improved by incorporating a coating additive. Several additives were evaluated and methylcellulose of high viscosity grade was found most satisfactory. The in vitro theophylline release was relatively linear and pH independent, and could be varied in a predictable manner by manipulating the coat thickness. In addition, when the coated pellets were subjected to additional thermal treatment, the drug release was stable after storage for one year [44].

4.3. REGULATION OF BIOCOMPATIBILITY OF POLYMERIC MATERIALS

The application of polymer devices in medicine, especially for their implantation in the body requires the analysis of biocompatibility of polymeric materials. In connection with this problem, the study of the adsorption of proteins on a nonphysiological surface is attracting the attention of many researchers.

The report of Griesser described a study using low pressure gas plasma techniques for the fabrication of surfaces designed for compatibility with biological tissue and fluids. The smoothness of the surfaces was investigated using scanning tunneling microscopy (STM) and the wettability was assessed by measurement of air/water contact angles. Plasma surface treatment rendered fluorocarbon polymers wettable by polar liquids, but the wettability decreased with time during storage of the samples in air. Plasma polymerization, on the other hand, allowed the deposition of thin coatings which were more stable with time. STM of plasma polymers, which were coated with a sputtered film of platinum or gold, showed that these coatings were smooth and continuous. The attachment and growth of human umbilical artery endothelial cells was used to assess the biocompatibility of several plasma polymer surfaces. There was some correlation between the ability of a surface to grow cells and the wettability of that surface with polar liquids, but significant differences in cell growth on surfaces with very similar hydrophilicity indicated that other factors are at least as important [45].

Analysis of products formed in the end-capping step during the synthesis of two block copolyurethanes being studied as biomedical materials showed the presence of dimeric soft segments and free diisocyanate. In this standard two-step synthesis, the presence of these compounds leads to block copolyether-urethane-ureas containing sizeable amounts of dimeric hard and soft segments. These standard copolymers were compared in terms of IR spectra, stress-strain properties, and dynamic mechanical properties to their pure analogs which contain no dimeric segments [46].

Studies on protein adsorption and platelet adhesion indicate that polymer surfaces can selectively adsorb proteins from whole blood in vivo, and the composition of this adsorbed protein layer influences the adhesion of platelets, and thus determines the thrombogenicity of the polymer. The microphase structure of the block copolyurethanes seems to often confer a selectivity in albumin adsorption, which appears to make these materials relatively nonthrombogenic. Analyses of the synthetic procedures for preparing these block copolyurethanes indicated that the standard two-step synthetic procedure introduces sizeable amounts of dimeric soft and hard segments into the copolymers. Fourier Transform IR-spectroscopy and ESCA were used to determine the bulk and surface morphology of these standard block copolyurethanes as well as their pure $(AB)_n$ analogs, so as to understand protein binding to their surfaces. A study of these materials as vascular grafts indicated that the bulk properties (i.e., the compliance of the graft) can also affect graft patency under in vivo conditions [47].

Surface design aimed at reduced adhesion and preserved functions of platelets is of great importance for extracorporeal devices. Matsuda and Ito [48] explored a coating technique using hydrophilic-hydrophobic block copolymers on a hydrophobic poly(acrylonitrile) (PAN) haemodialyser. The hydrophilic block of copolymers was composed of either poly(methoxy polyethylene glycol methacrylate) or poly(dimethyl acrylamide), and the hydrophobic block was poly(methyl methacrylate). The copolymers were coated on the dialyser membrane by means of a solution coating method. Upon coating, the hydrophobic block of the copolymers was anchored on a PAN membrane and the hydrophilic block oriented towards the blood-material interface. This was deduced from water wettability measurements. Significantly reduced transmembrane stimulation of platelets was observed, which was evaluated by determining the intracellular calcium ion concentration of platelets eluted through treated hollow fibers. This suppression was enhanced as the relative fraction of the hydrophilic block of the copolymers increased. Furthermore, the number of platelets adhering to the copolymer-coated PAN membrane was drastically reduced. Thus, coating of the hydrophilic-hydrophilic block copolymers provided better biocompatibility on a hydrophobic PAN dialyser [48].

Materials with the unique complex of the properties of elasticity and tensile strength processed from the natural rubber latex (NRL) were modified to improve their blood compatibility. ESCA, ATR-IR, UV-spectroscopy and other methods were used for the analysis of modification, structure, and properties of material. Thromboresistancy was studied in in vivo and ex vivo tests. It was shown that blood compatibility is raised after purification of the material from nonrubber components by means of two-stage extraction changing the physical-chemical parameters of the surface. Two different principles were used in the process of modification: (1) coating by thin polyurethane (PU) coverings with improved thromboresistant properties—thus the problem of providing high adhesion interaction of covering with the latex base was solved; (2) heparin surface immobilization—higher efficiency of modification of latex material in gel-form without preliminary protein adsorption was shown. Modification allowed for increased blood compatibility of the latex materials preserving at the same time their elasticity and tensile strength [49].

Currently, it is well known that the adsorption of plasma proteins is the primary act of thrombus formation on the polymer surface in contact with blood [50]. The most convincing version appears to be that the activation of platelets on the proteinated surface is stimulated by the partial rearrangement in structure of the protein which adsorbed on the surface [51, 52]. Earlier we showed [52–55] that the mechanism of adsorption constitutes a complex multistage process including diffusion of the protein globules to the polymer surface and the subsequent stages of formation of structurally rearranged and native protein layer. The correlation between the kinetic parameters of adsorption and the biocompatibility of polymers has been established either for materials giving a good [56] or unsuitable [57] account of themselves in medical practice. Then the approach was extended to test and alter the biocompatibility of materials widely used in medicine, particularly for design of drug-delivery systems. These are segmented polyetherurethanes and siloxanecarbonates [58].

To provide this purpose the diffusion-kinetic model was modified to be relevant to the process of adsorption onto the surface of block-copolymers, and FTIR-spectroscopy with accessories of attenuated total reflection (FTIR-ATR) was applied to express evaluation of the amount of adsorbed protein.

We have shown previously [52–55] that the equilibrium protein layer on the surface of hydrophobic and moderately hydrophilic polymers incorporates two adsorption layers of different structure, architecture, and intramolecular bond energy. The first (termed irreversible, or firmly adsorbed) consists of molecules subjected to structural rearrangement and is monomolecular; this layer interacts directly with active centers of the surface. The structure of the second (reversible) layer is close to the structure of native protein. Macromolecules contained on this layer are adsorbed by active centers of molecules which have undergone structural rearrangement. Interaction of native globules are neglected because of the absence of such interaction under normal conditions in solution. The formation of an irreversible layer onto the surface of block-copolymer may be presented in the form of second- and first-order parallel and step-by-step reactions

$$A_v + \Theta_{11} \underset{k_{121}}{\overset{k_{111}}{\rightleftharpoons}} A_{s11} \overset{k_{11}^*}{\longrightarrow} A_{s11}^*$$

$$A_v + \Theta_{12} \underset{k_{122}}{\overset{k_{112}}{\rightleftharpoons}} A_{s12} \overset{k_{12}^*}{\longrightarrow} A_{s12}^*$$

(4.19)

where A_v is the molecule in solution coming into contact with the surface; Θ_{1i} is the active center of the surface corresponding to the i-block; A_{s1i} is the native molecule which either undergoes structural rearrangement, or desorbs into the solution; A_{s1i}^* is the molecule that sustained structural rearrangement.

Similarly, a second-order equation may describe the adsorption of reversible layer

$$A_v + \Theta_{21} \underset{k_{221}}{\overset{k_{211}}{\rightleftharpoons}} A_{s21}$$

$$A_v + \Theta_{22} \underset{k_{222}}{\overset{k_{212}}{\rightleftharpoons}} A_{s22}$$

(4.20)

where Θ_{2i} is the active center on the surface of the corresponding layer, A_{s2i} is the molecule incorporated in the reversible layer.

The system of kinetic equations corresponding to the scheme (4.19–4.20) takes the form

$$dC_{s1}/dt = \sum_{i=1}^{2} k_{11i}C_v (0,t)(C_{s1i}^{\infty} - C_{s1i} - C_{s1i}^*) - (k_{12i} + k_{1i}^*)C_{s1i} \quad (4.21)$$

$$C_{s1} = \sum_{i=1}^{2} C_{s1i} + C_{s1i}^* \quad (4.22)$$

$$dC_{s1i}^*/dt = k_{1i}^* C_{s1i} \quad (4.23)$$

$$dC_{s2}/dt = \sum_{i=1}^{2} k_{21i}C_v (0,t)(N_iC_{s1i}^* - C_{s21i}) - k_{22i}C_{s2i} \quad (4.24)$$

$$C_{s2} = \sum_{i=1}^{2} C_{s2i} + C_{s2i}^* \quad (4.25)$$

Here C_{s1} is the overall concentration directly on the surface; C_{s1i}, C_{s1i}^*, and C_v (0,t) are the concentrations of molecules A_{s1i}, A_{s1i}^*, and A_v, respectively (surface coordinate $x=0$); C_{s1i}^{∞} is the maximum concentration of irreversible layer corresponding to protein adsorption by all active centers of i-sort of the surface; C_{s2i} is the concentration of molecules A_{s2i}; N_i is the average number of active centers of Θ_{2i}; k_{ijk} (i, j, k = 1,2) are the rate constants.

We have shown previously that the rate of adsorption of proteins depends not only on polymer-globular interactions, but also on hydrodynamic conditions of the experiment [53, 54, 56]. This is explained by the fact that concentration $C_v(0,t)$ is determined by the transfer in the diffusion boundary layer near the surface. According to the diffusion-kinetic model [54], protein concentration in a diffusion layer of a thickness l conforms to the equation

$$\delta C_v/\delta t = D\delta^2 C_v/\delta x^2 \quad (4.26)$$

with boundary conditions

$$-\bar{D}\delta C_v/\delta x = RdC_v/dt \qquad x = l$$

$$D\delta C_v/\delta x = dC_s/dt \qquad x = 0 \quad (4.27)$$

$$C_s = \sum_{i=1}^{2} C_{s1i} + C_{s1i}^* + C_{s2i}$$

where $\bar{D}$ is the diffusion coefficient of protein in solution; C_v is the protein concentration in the zone of mixing; $C_s = C_{s1} + C_{s2}$; R is the typical size of an experimental cell.

Constants in the system of Eqs. (4.21–4.27) may be divided into two groups. One group (rate constants, N, C^{oo}_{s1}) directly characterizes the polymer-protein system, on which properties depend, the second (1, C_v, R) is determined by experimental conditions and is given by a researcher. Coefficient D is a characteristic of a given protein and may be found from independent experiments. Controlling adsorption by changing flow rate (and therefore l) and the initial concentration of protein in solution, the necessary data may be obtained for calculating rate constants, N, and C^{oo}_{s1} and the prediction of properties of the polymer-protein system as a whole.

The general consequences from the model which are essential for further analysis of particular polymers are that

—all kinetic curves of irreversible adsorption tend to the limit (C^{oo}_{s1}) which is altered by the ratio of monomers (m) (Fig. 4.13) and

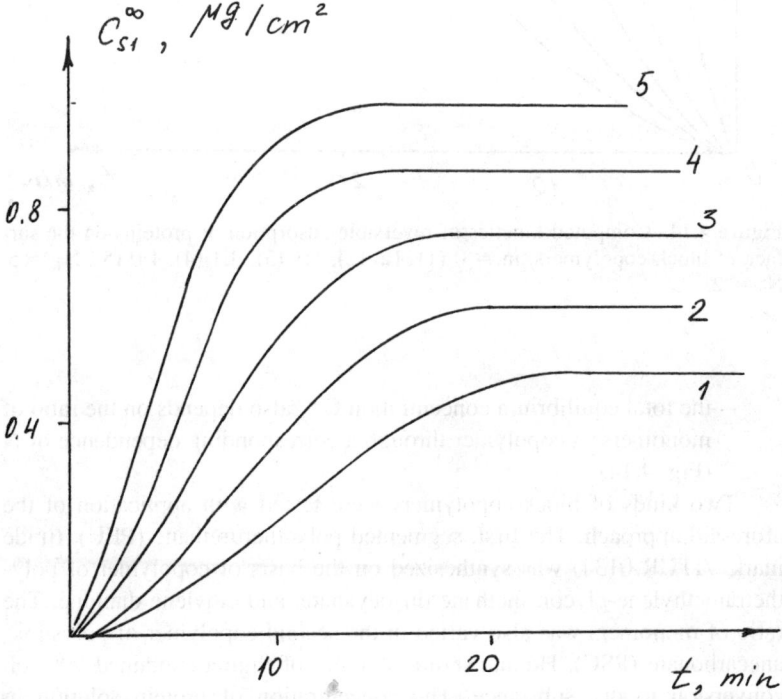

Figure 4.13. Computed kinetics of irreversible adsorption of proteins on the surface of block-copolymers. m = 0 (1), 1:3 (2), 1:1 (3), 3:1 (4), 1:0 (5). $C_{s11}^{oo} = 0.5$, $C_{s12}^{oo} = 1$.

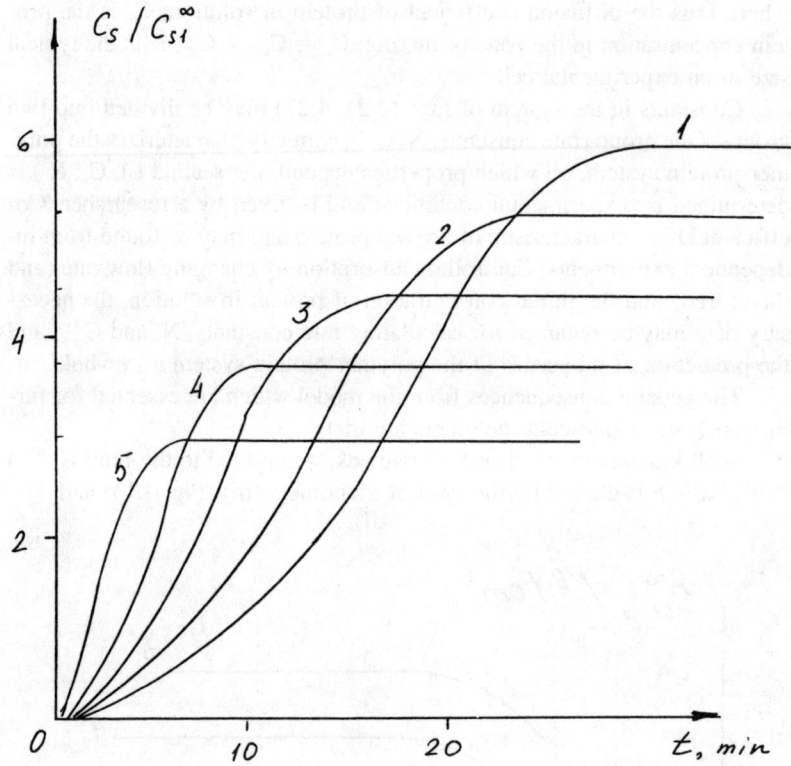

Figure 4.14. Computed kinetics of reversible adsorption of proteins on the surface of block-copolymers. m = 0 (1), 1:3 (2), 1:1 (3), 3:1 (4), 1:0 (5). $N_1 = 5$, $N_2 = 2$.

—the total equilibrium concentration $C_s^{\infty\infty}$ also depends on the ratio of monomers in copolymer through a corresponding dependence of N (Fig. 4.14).

Two kinds of block-copolymers were tested with application of the aforesaid approach. The first, segmented polyetherurethane (PEU), (trade mark VITUR-0134) was synthesized on the basis of copolymer of polythetramethylene-glycol, methane diisocyanate, and ethylene diamine. The ratio of monomers was also varied in the second copolymer of polysiloxanecarbonate (PSC). Human serum albumin of Sigma contained 99% on conversion to dry substance. The concentration of protein solution in Tirade-3 buffer (pH = 7.4) varied within the range of 0.01–5 g/l. Other

essential parameters of proteins should be indicated as 69000 of molecular weight and 6×10^{-7} cm^2/s of diffusion coefficient in buffer solution.
Special experimental order is required for accurate measurements of parameters involved in the model (4.21–4.27).

Parameters of reversible adsorption must be measured first from the
desorption experiments. The kinetics of desorption of a reversible layer
from the surfaces of polymers was examined at various rates of solvent
flow Q, initial bulk concentrations of protein, and different monomer ratio
in block-copolymers (as it is shown in Fig. 4.15, for example, of polysiloxanecarbonate). It was found that the total time of desorption relates to
a certain Q value as it follows the diffusion-kinetic model [57]. All kinetic
curves tend to one limit which corresponds to the concentration of irreversible layer under given ratio of monomers. In the meantime, the variation of this ratio alters such a limit. The examination of kinetic curves at
different protein concentrations in the initial solution showed that each
curve is characterized by its own equilibrium value of C_s, which also depends on the ratio of monomers in the block-copolymer. The rate constants

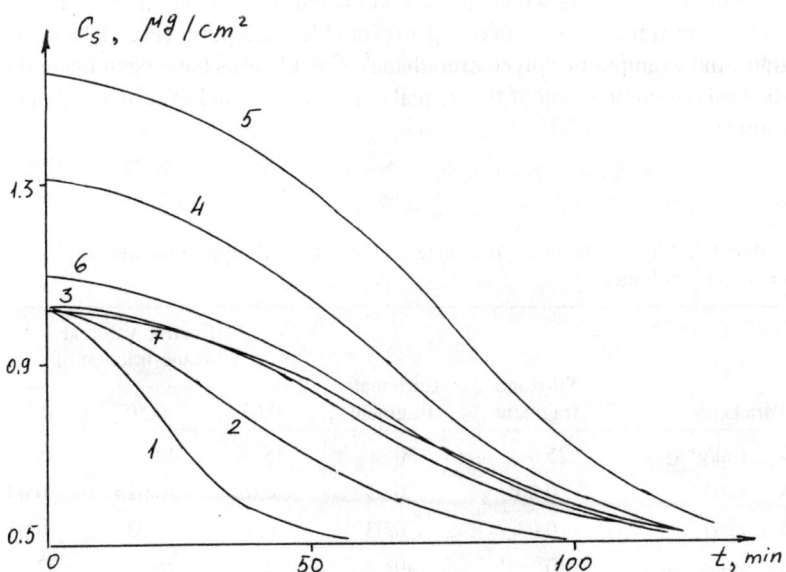

Figure 4.15. Kinetics of desorption of albumin from the PSC surface:
Q = 10 (1, 4–7), 30 (2), 60 (3) ml/min.; C_v^0 = 1 (1), 2 (4); 5 (5) g/l; m = 3:1 (1),
1:1 (6), 1:3 (7).

Table 4.6. Physicochemical parameters of albumin adsorption on the surface of segmented polyetherurethane.

Parameter	Hard segment	Soft Segment	Effective value at soft/hard segment ratio of		
			72/28	60/40	56/44
k_{11}, (ml/g*s)	50	30	35	40	45
k_{12}, (1/s)	0.05	0.01	0.02	0.25	0.03
k^*_1, (1/s)	0.03	0.03	0.03	0.03	0.03
k_{21}, (ml/g*s)	30	40	40	40	35
k_{22}, (1/s)	0.4	0.1	0.1	0.3	0.35
N	2	4	3.8	3	2.6
C^{oo}_{s1}, (mkg/cm)	0.8	0.6	0.6	0.65	0.7

of the reversible protein layer were calculated on the basis of the desorption kinetics, and results are presented in Tables 4.6 and 4.7.

Other parameters which are also collected in the table have been calculated from the kinetic curves of irreversible adsorption (Fig. 4.16 illustrates the example of polyetherurethane). Calculations have been made on the basis of comparison of theoretical curves (lines) and experimental data (points).

Table 4.7. Physicochemical parameters of albumin adsorption on siloxane-containing polymers

Parameter	Siloxane fragment	Carbonate fragment	Effective value at siloxane fraction of		
			0.65	0.50	0.25
k_{11}, (ml/g*s)	25	30	35	40	45
k_{12}, (1/s)	0.001	0.05	0.02	0.025	0.03
k^*_1, (1/s)	0.03	0.03	0.03	0.03	0.03
k_{21}, (ml/g*s)	10	40	22	25	32
k_{22}, (1/s)	0.014	0.1	0.04	0.06	0.08
N	4	2	3.5	3	2.8
C^{oo}_{s1}, (mkg/cm)	0.5	0.6	0.6	0.65	0.7

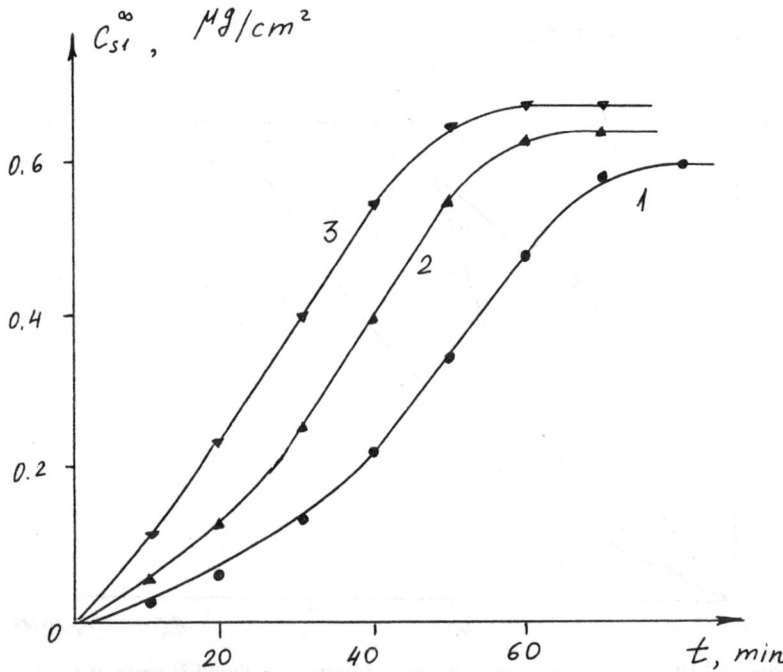

Figure 4.16. Kinetics of irreversible adsorption of albumin on the PEU surface: $Q = 10$ ml/min.; $C_v^0 = 1$ g/l; m = 3:1 (1), 1.5:1 (2), 1.25:1 (3).

The tables also contain the effective parameters which could be calculated using a previous unmodified model [52]. The validity of the modified approach was then checked by the comparison of theoretical and experimental results of overall albumin adsorption on the surface of polyetherurethanes with different ratio of hard and soft segments (Table 4.6, Fig. 4.17). Because of the good coincidence of theoretical and experimental data we can draw the following general conclusions. The increase of the soft segment in the chemical structure of polyurethanes causes the decrease of the total amount of irreversibly adsorbed protein accompanied by the shift of the reaction of formation of the reversible layer to adsorption. Although the number of reversible binding centers slightly depends on the kind of segment, the contribution of this number itself in the thermodynamics and kinetics of adsorption is negligible in comparison with the chemical structure of these centers following structural rearrangement of macromolecules of the corresponding irreversible layer. In other words, the effect of the kind of segment on the values of rate constants is much

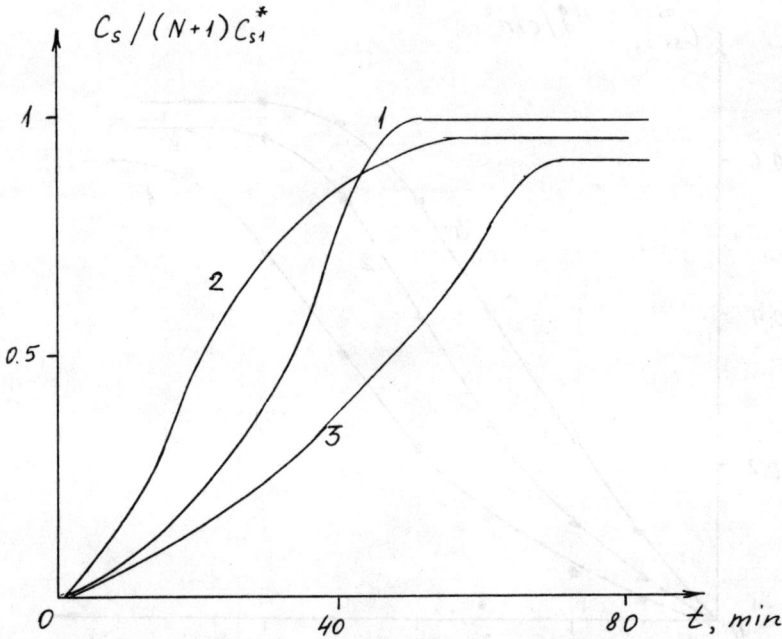

Figure 4.17. Kinetics of overall albumin adsorption on the PEU surface:
$l = 100$ μ; $C_v^0 = 45$ g/l; m = 72:28 (1), 60:40 (2), 56:44 (3).

more essential. The deactivation of the centers of the reversible adsorption
before the moment of contact between platelets and proteinated surface im-
proves the biocompatibility of the polymer. The worsening of the biocompatibility of the polymer coincides with the shift of equilibrium in the
reaction of formation of the reversible layer towards desorption.

The alteration of the polymer's biocompatibility by the controlled variation of the ratio between monomers was also shown for a case of albu-
min-polysiloxanecarbonate system (Table 4.7, Fig. 4.18). The results
illustrate the expected effect of the siloxane fraction of the copolymer on
the quantity of irreversibly adsorbed protein. The most considerable effect
the ratio of copolymer components has is on the reversible adsorption kinetics. The larger extent of structural rearrangement in proteins stimulates
protein-protein interactions. This intensifies the reversible adsorption and
increases the equilibrium constant of the reversible layer. Once this opposite effect of structural rearrangement on kinetics of adsorption exists, an
optimal ratio of monomers should be expected from the point of polymer

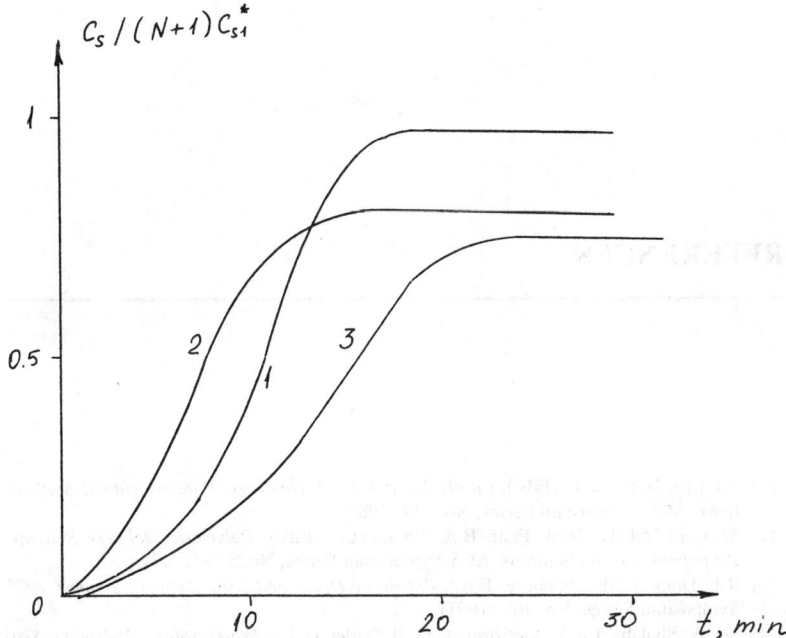

Figure 4.18. Kinetics of overall albumin adsorption on the PSC surface: $l = 100\ \mu$; $C_v^0 = 45$ g/l; m = 75:25 (1), 50:50 (2), 25:75 (3).

biocompatibility. For in vitro conditions this optimal ratio should be expected at value of 50/50 for polysiloxanecarbonate and 60/40 for polyetherurethane.

Therefore, the above diffusion-kinetic model allows the reversibility factor to be controlled with the time. This opens a way to improve biocompatibility of copolymers used in medicine.

REFERENCES

[1] P.I. Lee, W.R. Good (Eds.). *Controlled-Release Technology: Pharmaceutical Applica-tions,* ACS Symposium Series, No. 348 (1987).

[2] M.A. El-Nokaly, D.M. Piatt, B.A. Charpentier (Eds.). *Polymeric Delivery Systems: Properties and Applications.* ACS Symposium Series, No. 520 (1993).

[3] R.L. Dunn, R.M. Ottenbrite (Eds.). *Polymeric Drugs and Drug Delivery Systems.* ACS Symposium Series, No. 469 (1991).

[4] Sh. W. Shalaby, Ch. L. McCormick, G. B. Butler, (Eds.). *Water-Soluble Polymers: Syn-thesis, Solution Properties, and Applications,* ACS Symposium Series, No. 467 (1991).

[5] N.A. Plate, L.I. Valuev. *Polymer Contact with the Living Body.* Issue 8. Znanie, Moscow (1987). (In Russian).

6] A.L. Iordanskii, T.E. Rudakova, G.E. Zaikov. *Interaction of Polymers with Bioactive and Corrosive Media,* VSP, Utrecht, The Netherlands (1994).

[7] Gillespie M. *3i Technology Forecast. Issue Number 2. Drug delivery systems into 1990s.* 3i Group Plc. (1990).

[8] A.Ya. Polishchuk et al. *Polymer Sci. USSR,* **32A**, 2203 (1990).

[9] A.Ya. Polishchuk et al. *Proc. All-Union Conference on Pharmacology,* Kharkov, USSR (1989). (In Russian).

[10] K.L. Shantha, K.P. Rao. *J. Appl. Polym. Sci.,* **50**, 1863 (1993).

[11] R. Bowtell et al. *Magnetic Resonance Imaging,* **12**, 361 (1994).

[12] J. Rosenblatt, B. Devereux, D.G. Wallace. *Biomaterials,* **15**, 985 (1994).

[13] H.K. Sah, Y.W. Chien. *Drug Develop. Industr. Phar.,* **19**, 1243 (1993).

[14] J.H. Jou, P.T. Huang. *Polymer,* **33**, 1218 (1992).

[15] T.C. Dahl, I.I.T. Sue. *Pharm. Res.,* **9**, 398 (1992).

[16] S. Akhtar, C.W. Pouton, L.J. Notarianni. *J. Controlled Release,* **17**, 225 (1991).

[17] K. Petrak. *J. Bioactive and Compatible Polymers,* **8**, 178 (1993).

[18] R.S. Harland, N.A. Peppas. *Eur. J. Pharm. Biopharm.,* **39**, 229 (1993).

[19] R.S. Harland, N.A. Peppas. *J. Controlled Release,* **26**, 157 (1993).

[20] L.Y. Shieh, N.A. Peppas. *J. Appl. Polym. Sci.,* **42**,1579 (1991).

[21] H. Loth, A. Euschen. *Macromol. Chem. Rapid. Commun.,* **9**, 35 (1988).

[22] A.Ya. Polishchuk. *Intern. J. Polymeric Mater.,* **25**, 37 (1994).

[23] I.C. McNeill, A.Ya. Polishchuk, G.E. Zaikov. *Polymer Degrad. Stab.,* **37**, 223 (1992).

[24] A.Ya. Polishchuk et al. *Tekstil'naya Khimiya (Textile Chemistry),* **1**, 91 (1992). (In Rus-sian).

[25] Y. Okuyama et al. *J. Biomater. Sci. Polym. Ed.,* **4**, 545 (1993).
[26] R. Yoshida et al. *Industr. Eng. Che. Res.,* **31**, 2339 (1992).
[27] A.S. Hoffman, A. Affrassiabi, L.C. Dong. In: *IUPAC Macromolecules' 86 Meeting, Oxford,* p. 65, Butterworth-Heinemann Ltd. (1986).
[28] A.S. Hoffman, A. Affrassiabi, L.C. Dong. *J. Controlled Release,* **4**, 213 (1986).
[29] L.C. Dong, A.S. Hoffman. *J. Controlled Release,* **4**, 223 (1986).
[30] J. Verdu. *J. Macromol. Sci. Pure and Appl. Chem.,* **A31**, 1383 (1994).
[31] A. Michele, V. Vittoria. *Polymer,* **34**, 1898 (1993).
[32] M.H. Yang, T.J. Chu. *Polymer Testing,* **12**, 97 (1993).
[33] C.G.P.H. Schroen et al. *J. Membr. Sci.,* **80**, 265 (1993).
[34] G. Zingone, F. Rubessa. *STP Pharma Sci.,* **4**, 122 (1994).
[35] M.P. Patel, M. Braden. *Biomaterials,* **12**, 645 (1991).
[36] R.D. Toddywala et al. *Intern. J. Pharm.,* **76**, 77 (1991).
[37] Savilova L.B. et al. In: *Development of novel textile medical products,* Preprint of the Research Institute of Textile Materials, Moscow, 42 (1992). (In Russian).
[38] C.J. Kim, P.I. Lee. *Pharm. Res.,* **9**, 10 (1992).
[39] T.Z. Fu, C.J. Durning, H.M. Tong. *J. Appl. Polym. Sci.,* **43**, 709 (1991).
[40] U. Conte, L. Maggi, A. Lamanna. *STP Pharma Sci.,* **4**, 107 (1994).
[41] L.C. Li, Y.H. Tu. *Drug Develop. Industr. Phar.,* **17**, 2041 (1991).
[42] L.S.C. Wan, W.F. Lai. *Intern. J. Pharm.,* **81**, 75 (1992).
[43] R.S. Okor, S. Otimenyin, I. Ijen. *J. Controlled Release,* **16**, 349 (1991).
[44] K.H. Yuen, A.A. Deshmukh, J.M. Newton. *Drug Develop. Industr. Pharm.,* **19**, 855 (1993).
[45] H.J. Griesser et al. *Polymer Intern.,* **27**, 109 (1992).
[46] T.L.D. Wang, D.J. Lyman. *J. Polym. Sci. Polym. Chem.,* **31**, 1983 (1993).
[47] D.J. Lyman. *J. Bioactive and Compatible Polymers,* **6**, 283 (1991).
[48] T. Matsuda, S. Ito. *Biomaterials,* **15**, 417 (1994).
[49] N.V. Kislinovskaya, I.D. Khodzhaeva, S.P. Novikova, N.B. Dobrova. *Intern. J. Polymeric Mater.,* **17**, 131 (1992).
[50] G.E. Zaikov, A.L. Iordansky, V.S. Markin. *Diffusion of Electrolytes in Polymers,* VSP, Utrecht, The Netherlands (1988).
[51] S.D. Bruck. *Biomaterials,* **1**, 103 (1980).
[52] A.Ya. Polishchuk, A.L. Iordansky, G.E. Zaikov. *JMS-Rev. Macromol. Chem. Phys.,* **23C**, 33 (1983).
[53] A.Ya. Polishchuk, A.L. Iordansky, G.E. Zaikov. *Doklady (Reports) of the USSR Academy of Sciences,* **264**, 1431 (1982).
[54] A.Ya. Polishchuk, A.L. Iordansky, G.E. Zaikov. *Khimicheskaya Fizika (Chem. Phys.),* **1**, 1268 (1982).
[55] A.Ya. Polishchuk, A.L. Iordansky, G.E. Zaikov. *Polymer Science USSR,* **26**, 1068 (1984).
[56] A.Ya. Polishchuk et al. *Polymer Science USSR,* **27**, 1327 (1985).
[57] A.Ya. Polishchuk, A.L. Iordansky, G.E. Zaikov. *Kompositsionnye materialy (Composites),* **28**, 77 (1986).
[58] A.Ya. Polishchuk, G.E. Zaikov. *Proc. Intern. Conf. Rubber'94,* Moscow (1994).

5 APPLICATION OF POLYMER SYSTEMS FOR CONTROLLED RELEASE

This chapter includes a brief review of the application of polymeric systems for controlled delivery of low-molecular weight compounds in medicine, agriculture, and industry which were mostly made in early 1990s. Earlier examples were described fully in already quoted monographs (see references 1–7 of chapter 4). We also will not repeat the examples which were considered in the previous chapters.

5.1. SYSTEMS FOR CONTROLLED DELIVERY OF DRUGS

Systems of the controlled release of low-molecular solutes have found the main application in medicine, particularly to act as vehicles for delivering drugs to specific locations within the body. Petrak [1] has formulated two distinct approaches which are currently applied to improving the drug action via its mode of delivery:

(a) Controlled drug release, which aims to reduce or eliminate side effects by producing a steady therapeutic concentration of drug in the body. This is achieved through zero-order release of active drugs and does not usually lead to a changed distribution of the drug in the body. Examples of such systems are the transdermal and the osmotic pump deliveries.

(b) Site-selective drug delivery, which aims to ensure that the drug is delivered to the site of its biochemical and disease-related site of action, at the same time maintaining the drug inactive elsewhere in the body.

Controlled release systems exist in many forms, including specially designed tablets for oral administration, injectable microspheres, liposomes, or implants [2]. They may be also separated into nondegradable, water-soluble, or biodegradable. This is a general classification which we follow in the present chapter. There are many polymers which have been used as physical matrices for controlled delivery of drugs. To readers who are interested in examples of polymers from each class that have been used as drug delivery matrices and the criteria for their selection, we recommend the recent reviews of Dunn [3] and Jacobs and Mason [4]. This is, of course, apart from the above mentioned books.

5.1.1. Oral forms

Oral administration predominates in the systems for drug delivery for many reasons, the main of which is its simplicity. Preliminary sterilization is not generally required for such devices, and the risk of damage to the device is minimal. In the meantime, such a form requires regulation of hydrolytic stability and solubility of polymeric materials [5].

After oral administration of basic drugs, the different pH values of the gastrointestinal tract can result in drastic changes in drug solubility, which can be very high at acidic pH values and dramatically low at neutral/basic pH, with consequent problems for the design of oral extended release formulations. Giunchedi et al. [6] prepared an oral extended release formulation containing a basic drug, using dipyridamole as model. Three-component modified release granules capable of moderating the drastic dissolution behavior of dipyridamole were prepared by loading a swellable polymer (cross-linked sodium carboxymethylcellulose) with both the drug and an enteric polymer (cellulose acetate phthalate or cellulose acetate trimellitate). In vitro dissolution tests of modified release granules in USP gastric fluid showed a modulation of the high dissolution rate of dipyridamole at the acidic pH values, while in USP intestinal fluid a very marked improvement in drug dissolution was observed. Hydrophilic matrices containing the drug with the smoothed dissolution rate characteristics were prepared via mixing the granules with a gelling polymer (cellulose ether) and then tabletting the resulting mixture. In vitro release tests performed both at constant pH and with pH variation showed that the matrices were capable of providing extended drug release in both acidic and neutral/basic media [6].

The polymer coat may serve as a protector for a drug in the neutral oral cavity (pH = 6.8) and/or, for instance, acid medium of the stomach (pH = 2.0). A pH-sensitive system for release of caffeine from the swelling-controlled system was suggested in [7]. A copolymer of methylmethacrylate with N,N-dimethylaminoethylmethacrylate (MMA/DEAEA) with a monomer ratio of 7:3 was used as a polymer matrix. Copolymer disks of 1.3 mm in diameter and 0.33 mm in thickness contained 0.08 g of caffeine per gram of matrix. The effect of pH on the rate of release of caffeine from the MMA/DEAEA copolymer is shown in Fig. 5.1. The introduction of special air bubbles prolonged the effect of the drug in comparison with a tablet with quindine gluconate as a filler [8].

Hydroxypropylmethylcellulose (HPMC) became very popular in the formulation of controlled release tablets, mainly because of its hydrophilic and swelling properties. In HPMC tablet matrix systems, the drug release occurs mainly by Fickian diffusion and by polymer relaxation. Vigoreaux and Ghaly proposed a model to quantify the amount of drug released by these two phenomena. [9]. Their recent studies showed that it is possible to modify the kinetics of drug release by restricting matrix swelling. Results obtained gave evidence of a direct relationship between releasing areas and the amount of drug dissolved. Tablets with matrix swelling re-

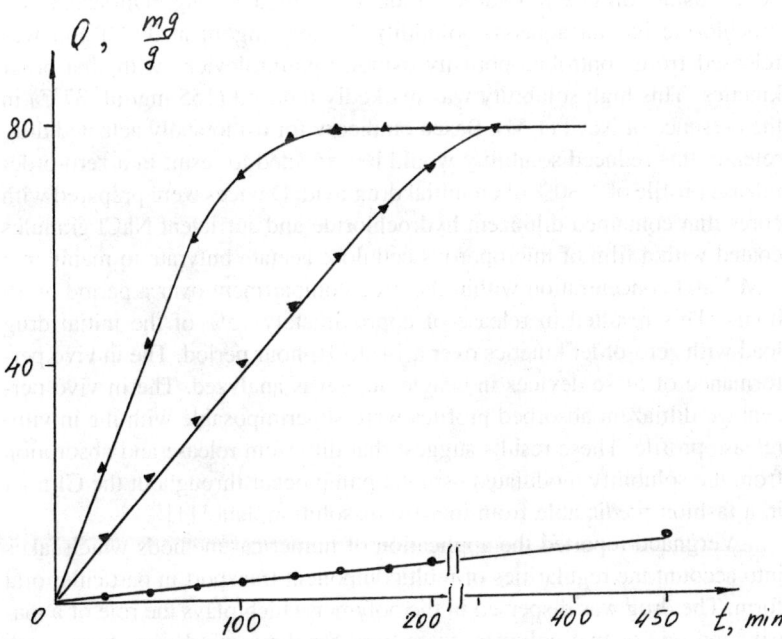

Figure 5.1. Release of caffeine from MMA/DEAEMA copolymer at pH = 7 (1), 5 (2), and 3 (3).

strictions exhibited a shift towards drug release by relaxational mechanism, which makes this technique a useful tool when a shift towards constant drug release is desired.

The drug release kinetics from controlled porosity osmotic pumps were effectively manipulated through application of either solubility- or resin-modulation methods. The solubility of diltiazem hydrochloride was reduced for an extended period of 12–14 hours through incorporation of controlled release sodium chloride elements into the core tablet formulations. Other diltiazem hydrochloride core tablets were prepared which contained the positively charged anion-exchange resin (poly(4-vinylpyridine)). In both instances, in vitro diltiazem hydrochloride release profiles that were zero-order and pH-independent were obtained without chemical modification of the drug. Release from devices that contained neither solubility- nor resin-modulation components was substantially first-order and highly pH-dependent. Solubility-modulated devices administered to dogs released diltiazem hydrochloride with similar in vivo and in vitro kinetics [10].

A generalized method was investigated for conversion of controlled-porosity osmotic pump release profiles from first-order to zero-order kinetics using diltiazem hydrochloride as a model drug. Diltiazem hydrochloride has an aqueous solubility (> 590 mg/ml at 37°C) and was released from controlled-porosity osmotic pump devices with first-order kinetics. This high solubility was markedly reduced (155 mg/ml; 37°C) in the presence of NaCl (1 M). Based on theory for osmotically actuated drug release, this reduced solubility would be expected to result in a zero-order release profile of >80% of an initial drug load. Devices were prepared with cores that contained diltiazem hydrochloride and sufficient NaCl granules coated with a film of microporous cellulose acetate butyrate to maintain a 1 M NaCl concentration within the drug compartment over a period of 16 hours. This resulted in release of approximately 75% of the initial drug load with zero-order kinetics over a 14- to 16-hour period. The in vivo performance of these devices in beagle dogs was analyzed. The in vivo percentage diltiazem absorbed profiles were superimposable with the in vitro release profile. These results suggest that diltiazem release and absorption from the solubility modulated osmotic pump occur throughout the GI tract in a fashion predictable from in vitro dissolution data [11].

Vergnaud reported the application of numerical methods which takes into account the regularities of multicomponent transport in particular oral form. The drug was dispersed in the polymer which plays the role of a matrix, and one or two polymers were used for the same dosage form, each polymer conferring a particular property. The process of release of the drug was rather complex. The gastric (or intestinal) liquid entered the polymer, dissolved the drug, and thus enabled the drug to leave the dosage form through the liquid located in the polymer. These two matter transfers (liquid and drug) were controlled by transient diffusion. Moreover, the diffusivity of the drug was generally expressed in terms of the concentration of the liquid in the polymer. Numerical approach was able not only to predict the kinetics of release of the drug, but also to determine the concentration profiles of both the liquid and the drug, in the dosage form [12].

Cellulose acetate (CA) latex plasticized with 150% triacetin (TA) and 120% triethylcitrate (TEC), based on polymer weight, provided dense and homogeneous films when deposited onto propranolol HCl tablets using conventional fluid bed technology. Film permeability to the drug was low, and flux-permeability enhancers were added to the CA structure during its manufacture. Films containing 40% sucrose and 10% PEG 8000 were found to provide the best release characteristics in terms of small lag-time (1 hour) and drug release profile (over 12 hours). When sucrose was added to TA or TEC plasticized films, a macroporous membrane was created during exposure to the dissolution fluid due to sucrose release from the film. These observations were consistent with the controlled porosity walls previously described for CA films deposited from organic solvents. Kelbert and Bechard postulated that drug mass transport occurs mainly within the porous CA structure, and the mechanism responsible for it is a combination of molecular diffusion/osmotic pressure via water transport into the porous cellulose acetate membrane [13].

The release of caffeine and ibuprofen as model drugs from a core-in-cup oral drug delivery system was used by Danckwerts [14] in order to determine their time exponent t^n against release profiles. The core-in-cup drug delivery system consisted of cores of various concentrations of two grades of hydroxypropylmethylcellulose (HPMC). HPMC K-4M and HPMC K-15M were the polymers in the core matrix. The flat disc-shaped core was then compressed with a previously compressed cup-shape tablet consisting of inert and impermeable carnauba wax and ethylcellulose. These drug delivery systems released active drug at a zero-order rate for periods of time between 8 and 23 hours. The release rate was then compared to core only systems. The t^n exponents varied from 0.477 for the lowest core system to 0.997 for a 5% w/w HPMC K-4M in ibuprofen core-in-cup system. Because the core, the cup, and the core-in-cup were compressed on an automated tabletting press, this drug delivery system could be easily scaled up to commercial production.

A hydrophobic material and surfactant coating for an oral dosage form, which allows for rapid drug release after a predetermined lag-time, was described by Pozzi et al. [15]. Their pulsed release system had the advantage of being manufactured using conventional film coating techniques and used excipients normally present in pharmaceutical formulations. The in vivo behavior of the device was evaluated in three scintigraphic studies. The lag-time for the system was found to be independent of normal physiological conditions, such as pH, digestive state of the subject, and the anatomical position at the time of release. At the end of the lag time, disaggregation of the core was both rapid and complete. Scintigraphic evaluation was also used to establish in vitro methodology capable of predicting the subsequent in vivo performance of the TIME CLOCK system. Pharmacokinetic studies using salbutamol as a model drug demonstrated that drug absorption was not affected by the in vivo behavior of the TIME CLOCK system.

The feasibility of preparing oral sustained release suspensions by using sparingly soluble salts of soluble ionic drugs was assessed in the investigation of Shah and Chafetz [16]. They used diltiazem as a model drug and cellulose acetate butyrate (CAB) as the coating polymer. A less soluble pectate salt of the drug was prepared and encapsulated by a solvent evaporation process. Percent drug release from CAB microcapsules was modified by using hydrophilic polymers and varying microcapsule drug load. The release was independent of pH and ionic strength of the dissolution medium, and the profiles fit a bi-exponential equation, suggesting a biphasic release mechanism. An initial burst effect was obtained followed by slow diffusion of drug through the polymeric coat. Suspensions of diltiazem pectate-loaded microcapsules were prepared in a preserved medium containing sorbitol, syrup, and methylcellulose. Redispersibility of suspensions was satisfactory at room temperature and 40°C, but poor at 37°C. The suspended microcapsules resulted in 8–12% increase in drug release after 1 week storage at room temperature compared to dry microcapsules; however, the release did not increase any further upon extended storage. The suspensions were unstable at 37°C, but remained relatively stable at or below room temperature for up to 26 weeks [16].

Sustained release phenylpropanolamine hydrochloride (PPH) granules and tablets were prepared using HPMC, HPMC and SCMC, Eudragit RS, Eudragit RS+L, or HPMC + Eudragit RS matrices. The release pattern of PPH from the prepared granules and tablets was found to be in the following order HPMC > HPMC + SCMC > RS > RS + 1 > HHPMC + RS. The results revealed that, although the drug concentration was kept constant in all the prepared granules and tablets, the drug release from these formulations was clearly different and depends mainly on the type of matrix used. The presence of Eudragit L with Eudragit RS and Eudragit RS with HPMC resulted in a marked decrease in the drug release compared with that obtained from the matrix containing HPMC or Eudragit RS alone. The release data of PPH from the prepared granules and tablets were treated mathematically according to zero-order, first-order, Langenbucher, modified Langenbucher, and Higuchi models. The results revealed that no one model was able adequately to describe the drug-release profiles from these formulations. In vivo studies in human volunteers showed that the peak urinary excretion of PPH occurred over a sustained period from 2 to 6.5 hours in the case of HPMC + SCMC tablets and from 2 to 10 hours in the case of either RS + L or HPMC + RS tablets [17].

A pulsatile drug-release system with a dry-coated tablet containing pentoxifylline was investigated by Otsuka and Matsuda [18] for controlling drug release in the GI tract. The system consisted of a core tablet with disintegrator and outer layer, which obtained compression from the ground mixtures of pentoxifylline and behenic acid. Drug release from a dry-coated tablet was investigated at 37°C in JPXII 2-nd fluid at pH = 6.8. The drug release from the outer layer was fitted to the Cobby model. The drug release from the wax matrix increased significantly after tablet disintegration, and the drug-release profiles showed typical sigmoidal curves. The disintegration time depended on the weight fraction of the core tablet, and the drug-release rate after disintegration increased with increasing drug concentration in the core tablet. The relationship between the time required for 50% drug release and the disintegration time was linear, indicating that the drug-release rate was controlled by regulating the disintegration time.

The preparation of the compressed tablets containing guaifenesin (model drug), calcium acetate (reactant), and pharmaceutical excipients were reported in [19]. The tablets were coated with calcium alginate hydrogel using a novel, self-correcting membrane coating technique. Effects of coating time, the type of alginate polymer, and pH of the dissolution medium on the rate of drug release were evaluated. In distilled water, zero-order drug release profiles were obtained from the coated tablets. The release rate decreased with an increase in the coating time (increased coat thickness) and molecular weight of alginate polymer. The release rate constants correlated with the model for the spherical reservoir system and, were used to calculate permeability of guaifenesin in the calcium alginate coatings. Alginate polymer with higher guluronic acid content provided acid stable coating and, higher molecular weight polymer produced a membrane with lower permeability for guaifenesin [19].

Pilot investigations of a small-scale method for production of calcium alginate matrices containing the model drug theophylline were performed by Ostberg and Graffner [20]. Gel beads of calcium alginate were produced by dripping a sodium alginate solution containing dispersed drug into a calcium chloride bath. Subsequent drying produced small matrices in which the drug crystals were embedded. Various factors connected with droplet formation, gelation, washing and drying of the product were investigated and shown to influence the size, the composition, and the release properties of the matrices. The encapsulation method offered an effective incorporation of theophylline only when the gelling and washing solutions were saturated with drug. Increasing the calcium concentration used for gelation reduced the encapsulation efficiency and gave a lower drug content of the matrices. By decreasing the calcium concentration of the gelling solution from 0.20 M to 0.05 M, or by removing more of the unbound calcium from the gel beads by more extensive washing, a slower drug release from the matrices was induced. The drying method and the drying temperature determined the appearance, the moisture content, and the physical state of the drug in the matrices. The drug release was markedly more rapid from freeze-dried than from air-dried matrices [20].

Gel/micelle materials, comprising hydrogels containing block copolymer micelles, have the potential to be very useful in a variety of extraction and controlled release applications. Calvert et al. [21] demonstrated that block copolymer surfactants can be immobilized in calcium alginate gels and that the resulting composite material preferentially solubilizes the model hydrophobic solute naphthalene. Five different polyethylene oxide-polypropylene oxide-polyethylene oxide triblock copolymer surfactants were investigated, and the qualitatively different properties exhibited are interpreted in light of the properties of these surfactants in solution. Alginate gel concentrations of 22.5 and 52.5 g/l and surfactant concentrations ranging from 20.4 to 136 g/l were considered. The data indicated in [21] showed that micelles can be permanently immobilized for purposes of extraction or controlled release and that the extent of solute uptake or release can be altered by varying gel and surfactant concentration and the type of surfactant employed.

A multiparticulate sustained release formulation of theophylline was developed, evaluated in vitro, and reported in [22]. The formulation comprised spherical pellets of high drug loading, coated with a rate controlling membrane. The pellets were prepared using an extrusion spheronization method, while coating was performed with an aqueous dispersion of ethylcellulose using a fluidized bed coating technique. When ethylcellulose was used alone as the coating polymer, the drug-release profile was unsatisfactory, but could be improved by incorporating a coating additive. Several additives were evaluated and methylcellulose of high viscosity grade was found most satisfactory. The in vitro theophylline release was relatively linear and pH independent and could be varied in a predictable manner by manipulating the coat thickness. In addition, when the coated pellets were subjected to additional thermal treatment, the drug release was stable after storage for one year [22].

This system was then evaluated in vivo [23]. Two preparations that differed solely in the coat thickness were studied in comparison with a solution of the drug. Both preparations produced serum concentration profiles that were reflective of a slow and sustained rate of absorption. The in vivo release against time profiles, which was calculated using a deconvolution procedure, showed that the two preparations differed in the rate but not the extent of drug release. Satisfactory correlation was also obtained between the in vivo and the in vitro results. When the two preparations were further compared using the parameters, time to reach peak concentration (T_p), peak concentration (C_p), and total area under the serum concentration versus time curves (AUC), a statistically significant difference was observed in the T_p and C_p values but not the AUC values, suggesting that the preparations differed in the rate but not the extent of absorption. In addition, the extent of absorption from both preparations was comparable to that obtained with the drug solution [23]. Another example of a diffusion-controlled delivery system for theophylline was recently reported in [24].

Interpenetrating polymer networks (IPNs) based on polyacrylamide and poly (vinyl alcohol) for hydrogel capsules were synthesized by Ramaaj and Radhakrishnan [25]. These polymer networks were evaluated as drug-delivery devices using Crystal Violet and Bromothymol Blue as model drugs. The observed drug release was higher for semi-II-IPN than full-IPN. The drug-release behaviors from these capsules were analyzed by the exponent relation $M_t/M_{oo} = Kt^n$, where "K" and "n" are constants and M_t/M_{oo} is the fraction of the drug released until time "t." The constant "n" was found to be above 0.5, which suggests that the release of drug from the capsules follows the non-Fickian diffusional model [25].

Tablets of acetaminophen as a model drug were prepared with low-substituted hydroxypropylcellulose (L-HPC) of various particle sizes at various loading in the formulation. Drug release into an aqueous dissolution medium (pH 1.2) was sustained from tablets prepared with fine L-HPC (LH41) at loading of more than 20%. Tablets prepared with less than 20% LH41 or with coarse L-HPCs (LH11, LH21, and LH31) disintegrated in the medium, resulting in rapid release of the drug. The difference in behavior could not be explained in terms of differences in tablet strength, but in swelling and water uptake abilities of the tablet's polymer. Swelling work (swelling force), water penetration speed, and water uptake of LH41 (4.4-mμ average particle size) were much smaller than those of coarse L-HPCs. The formation of a continuous gel-like layer on the surface of tablets containing more than 20% LH41 was another factor to sustain the drug-release rate [26].

A controlled-release matrix filler was prepared by Kawashima et al. [27] using spray-drying a heated aqueous hydroxypropylmethylcellulose solution suspending microcrystalline cellulose (MCC, PH101). Acetaminophen tablets (used as model drug, content $= 50\%$) were prepared by directly compressing the mixture of drug and spray-dried matrix filler. When HPMC was formulated with more than 10% of the matrix filler, drug release from the tablets was satisfactorily sustained. To obtain a similar sustained-release pattern with unmodified original HPMC, more than 50% of the matrix filler was required in the formulation. Whereas, when the tablet formulation was less than 5% modified HPMC, the drug was rapidly released from the tablets. Uniformly distributed HPMC in the spray-dried filler should lead to such drug-releasing behaviors. The micrometric properties of HPMC in the matrix filler, such as particle size, size distribution, and the loading amount of HPMC, were main factors in determining the drug-release properties of the tablets. The drug-release rate of the tablets was determined by the erosion rate of the gelled HPMC formed on the surface of the tablets. The drug-release kinetics were described as a function of the cube root of the tablet weight. The drug was tabletted directly with the modified matrix filler by a rotary tabletting machine [27].

A new anionic composite bead system with a transient membrane-matrix structure, capable of prolonged constant-rate drug release, was developed by Lee from suspension-polymerized poly(methyl methacrylate-co-methacrylic acid) (PMMA/MAA). These composite beads have a thin PMMA/MAA surface layer and a core consisting of the sodium salt form of the polymer (PMMA/MANa). The high loading ($> 20\%$) of a model drug (oxprenolol HCl) that is achievable in this system from a loading solution concentration as low as 0.5% suggests the formation of a drug-polymer complex in the form of an ionic salt in the core. The release of oxprenolol from such composite beads shows an initial burst effect followed by an extended constant-rate region before leveling off. Apparently, the surface PMMA/MAA layer functions as a transient rate-controlling membrane before it is completely ionized. Because the ionization process was slow, the rate-controlling characteristics of the surface layer and the resulting constant rate of drug release were both sustained for an extended period. The unique feature of Lee's system is not only its high drug loading capability, but also the transient nature of the rate-controlling surface layer, which is completely ionized towards the latter part of the drug release, thus avoiding prolonged tailing of drug release that is normally associated with permanent membrane-matrix systems [28].

We shall give more examples of oral forms in section 5.1.3 where microcapsules and microspheres will be considered in greater detail.

Generally, oral administration is the route of choice in the daily practice of pharmacotherapy. However, in some circumstances this is impractical or even impossible (during nausea and vomiting or convulsions, in uncooperative patients, and before surgery). In these cases, as indicated in

[29], the rectal route may represent a practical alternative and rectal administration is now well accepted for delivering, for example, anticonvulsants, nonnarcotic and narcotic analgesics, theophylline, antiemetics, and antibacterial agents, and for inducing anesthesia in children. It may also represent an interesting alternative to intravenous or other injection routes of drug administration. The rate and extent of rectal drug absorption are often lower than with oral absorption, possibly an inherent factor owing to the relatively small surface area available for drug uptake. In addition, the composition of the rectal formulation (solid against liquid, nature of the suppository base) appears to be an important factor in the absorption process by determining the pattern of drug release. This relation between formulation and drug uptake has been clearly demonstrated for drugs like diazepam, paracetamol (acetaminophen), indomethacin, methadone, and diflunisal. Coadministration of absorption-promoting agents (surfactants, sodium salicylate, enamines) represents another approach towards manipulating rectal drug absorption, although this concept requires further research concerning both efficacy and safety. For a number of drugs, the extent of rectal absorption has been reported to exceed oral values, which may reflect partial avoidance of hepatic first-pass metabolism after rectal delivery. This phenomenon has been reported for morphine, metoclopramide, ergotamine, lidocaine (lignocaine), and propranolol. Rectal drug delivery in a site- and rate-controlled manner using osmotic pumps or hydrogel formulations may provide opportunities for manipulating systemic drug concentrations and drug effects. The extent of first-pass metabolism may be influenced (lidocaine), depending on the site of drug administration in the rectum. The rate of delivery may determine systemic drug action and side-effects (nifedipine), and it may affect the local action of concurrently administered absorption promoters on drug uptake (cefoxitin). Local irritation is increasingly being acknowledged as a possible complication of rectal drug therapy. Long-term medication with rectal ergotamine and acetylsalicylic acid, for example, may result in rectal ulceration, and irritation after a single administration of several drugs and formulations has been described. The assessment of tolerability and safety is imperative in the design of rectal formulations. Recent studies corroborate the clinical relevance of rectal drug therapy, and the value of the rectal route as an alternative to parenteral administration has been assessed for several drugs, e.g., diazepam, midazolam, morphine, and diclofenac. Further development and optimization of rectal drug formulations for clinical use may be expected in the near future. In addition, data demonstrate the applicability of rectal therapeutic systems, e.g., osmotic devices, in investigative pharmacokinetic and pharmacodynamic studies, and in assessment of the design specifications of controlled release formulations [29].

5.1.2. Transdermal and other on-skin devices

Many systematically active pharmaceuticals and biopharmaceuticals are reportedly subjected to extensive presystemic elimination when taken orally. To circumvent this dilemma in systemic delivery, extensive research efforts have recently been devoted to exploring various nonparenteral (noninvasive) routes of administration for enhancing systemic bioavailability of pharmaceuticals and biopharmaceuticals, which have been reportedly subjected to extensive presystemic elimination, via the bypassing of hepato-gastrointestinal "first-pass" metabolism. Using nicotine, a pharmaceutical, and enkephalin, a biopharmaceutical, the systemic delivery of pharmaceuticals and biopharmaceuticals through the skin and the various absorptive mucosae was illustrated by Chien and colleagues [30]. The mechanisms and kinetic processes involved for their systemic delivery and various technical issues were discussed in that article.

The skin permeation and release kinetics of nicotine from four nicotine-releasing transdermal delivery systems (TDS) marketed recently were investigated under identical conditions to evaluate the effect of system design and the interchangeability of these products. In the study, hairless rat skin was first used as an animal model to evaluate the permeation mechanisms of various TDSs, which were then verified by studying the permeation through human cadaver skin. Three of the four TDSs were found to deliver nicotine at zero-order permeation kinetics at steady state with permeation rate ranging from 0.072 to 0.197 $mg/(cm^2*hour)$, while the fourth one produced a triphasic zero-order permeation rate profile. Three TDSs released nicotine at nonlinear manner, which could be described by Fickian relationship, while one TDS yielded a constant release at steady state. The different skin permeation profiles of nicotine delivered by these TDSs could be explained by the difference in their system designs and structural compositions [31].

The key aspects of the pharmacokinetics of TDSs, including time lag, steady-state plasma levels, and decline phase were reviewed by Berner and John [32]. The 7 currently marketed transdermal systems [nitroglycerin (glyceryl trinitrate), estradiol, clonidine, fentanyl, nicotine, scopolamine (hyoscine), and estradiol/norethisterone acetate] were discussed, as were systems in development. Single-dose absolute bioavailability studies characterized the period of onset, the steady-state plateau and the declining phase, and typified transdermal delivery. Clinically, these systems were used to achieve multiple peak serum estradiol concentrations after application of transdermal estradiol, and an initial peak systemic concentration of testosterone after application of transdermal testosterone. Multiple-dose, dose proportionality, and skin site bioequivalence studies were needed for the full pharmacokinetic characterization of a TDS. Although the data were limited, population factors, cutaneous metabolism, and tolerance all appeared to influence the disposition of drugs administered transdermally. For example, the route of delivery influenced which nitroglycerin metabolite predominated. Furthermore, as a result of tolerance to nitrates, a TDS must be removed for 8 to 12 hours for optimal effect. Therefore, TDSs, designed on the basis of pharmacokinetic principles and concentration-effect relationships, have the potential to provide optimal therapy for the treatment of some conditions [32].

The transdermal drug delivery systems based on polymeric pseudolatex and matrix diffusion controlled systems for salbutamol were prepared and compared for in vitro skin permeation profile and in vivo performances. Poly (isobutylene) was used as release controlling polymer in both the systems. In vitro skin permeation was studied using the human cadavar skin in franz diffusion cell. Permeation rate constants for matrix diffusion controlled system and pseudolatices were 10.625 and 13.750 $\mu g/(hour*cm^2)$, respectively. The prepared transdermal systems were tested on human volunteers having chronic reversible airways obstruction and compared with oral treatments (Asthaline). The in vivo drug plasma profiles following transdermal and oral treatments reveal that although peak plasma level by oral administration was higher in comparison with the transdermal treatments, troughs and peaks were discernible at dosing times. In the case of transdermal treatments, constant drug plasma levels were recorded indicating controlled and systemic delivery of drug spaced over 30 hours. Among the prepared transdermal drug delivery systems, pseudolatices demonstrated better drug plasma profile, maintained at relatively higher level and flatter in appearance [33].

Iontophoresis enhances transdermal drug delivery by three mechanisms: (a) the ion-electric field interaction provides an additional force which drives ions through the skin; (b) flow of electric current increases permeability of skin; and (c) electroosmosis produces bulk motion of the solvent itself that carries ions or neutral species, with the solvent "stream." The relative importance of electroosmotic flow is the subject of the review [34]. Experimental observations and theoretical concepts are reviewed to clarify the nature of electroosmotic flow and to define the conditions under which electroosmotic flow is an important effect in transdermal iontophoresis. Electroosmotic flow is bulk fluid flow which occurs when a voltage difference is imposed across a charged membrane. Electroosmotic flow occurs in a wide variety of membranes, is always in the same direction as flow of counterions, and may either assist or hinder drug transport. Since both human skin and hairless mouse skin are negatively charged above about pH 4, counterions are positive ions and electroosmotic flow occurs from anode to cathode. Thus, anodic delivery is assisted by electroosmosis, but cathodic delivery is retarded. Water carried by ions as "hydration water" does not contribute significantly to electroosmotic flow. Rather electroosmotic flow is caused by an electrical volume force acting on the mobile counterions. The simple "limiting law" theory commonly given in textbooks and some research articles is a very poor approximation for transdermal systems. However, several extensions of the limiting law are compatible with each other and with the available experimental data. One of these theories, the Manning theory, has been incorporated into a theory for the effect of electroosmotic flow on iontophoresis, the latter theory being in good agreement with experiment. Both theory and experimental data indicate that electroosmotic flow increases in importance as the size of the drug ion increases. The "ionic" or Nernst-Planck effect is the largest contributor to flux enhancement for small ions. Increased skin permeability or the skin "damage effect," is a significant factor for both large and small ions, particularly for experiments at high current density. For monovalent ions with Stokes radii larger than about 1 nm, electroosmotic flow is the dominant flow mechanism. Because of electroosmotic flow, transdermal delivery of a large anion (or negatively charged protein) from the anode compartment can be more effective than delivery from the cathode compartment [34].

A mathematical model was presented in [35] for the description of transdermal drug delivery from a matrix-type delivery device. The model is partly diffusional and partly compartmental in nature. The matrix and stratum corneum were both considered to be diffusion layers, connected to a three-compartment model representing the viable epidermis/dermis, plasma, and peripheral tissues. The diffusion equation was solved numerically for the two diffusion layers under nonsink conditions. The ordinary differential equations for the compartmental model were also solved numerically. Combination of the two numerical solutions yielded a model which directly related the properties of the matrix to the profile of drug mass in the plasma and the urinary excretion profile. The model was first used to analyze data obtained from an in vivo trial of a matrix-type transdermal delivery device for the drug clenbuterol. Fitting of the model to the profile of drug concentration in the plasma, the urinary excretion profile, and the mass of drug remaining in the matrix with a modified simplex method yielded values for the model constants. These compared very favorably with independent values taken from the literature. Simulations of the influences of drug diffusivity within the stratum corneum, drug loading in the matrix, matrix thickness, and drug diffusivity within the matrix on the profile of drug concentration in the plasma were then made. The model was not restricted to a steady state nor did it specify particular drug-release kinetics from the matrix. It did assume isotropic diffusion layers and spontaneous partitioning at boundaries [35].

An adhesive polymer drug dispersion-type transdermal drug delivery (a-TDD) system, consisting of a single drug-loaded adhesive polymer layer sandwiched between a drug-impermeable backing membrane and a detachable release liner, was developed from two silicone-based pressure-sensitive adhesive polymers for the controlled administration of drugs. The effect of variation in penetrant lipophilicity, using a series of testosterone derivatives with an increasing number of methyl groups in the steroid skeleton, on the release kinetics from the a-TDD system and skin permeation rate profiles was investigated. Absence of a methyl group at the enzymes present in the viable epidermis and dermis contributes to a wide variety of extrahepatic metabolism. During the course of this investigation, Toddywala and Chien found that the androgen esters they used for the study were completely bioconverted in the skin [36].

The in vitro release of model drugs, with a wide range of aqueous solubility, from monoolein-water liquid crystalline matrix systems was investigated by Burrows et al. [37]. Release of melatonin, pindolol, propranolol, and pyrimethamine from individual systems with initial drug loading concentrations within the range 1–20% w/w and atenolol from systems at concentrations up to 10% w/w could be fitted to both diffusion-controlled or first-order kinetics. The release of atenolol at initial drug loading concentrations of 15 and 20% w/w could be fitted to a zero-order release model. Release rates were related to the solubility of the drugs in the monoolein-water systems. Changes in the matrix monoolein/water weight ratio over the range 4:1–1:1 had no significant influence on drug release. Monoolein-water-drug systems prepared using drugs with either a high solubility (propranolol) or a low solubility (pyrimethamine) were stable when stored in the dark at 4°C for up to 6 months with no significant change in release characteristics. Systems incorporating propranolol were unstable when stored at 26°C for 15 days; storage of systems incorporating pyrimethamine under the same conditions were stable with no change in release characteristics [37].

5.1.3. Microcapsules and microspheres

Nanotechnology found remarkable application in the development of systems for drug delivery and the production of microspheres and microcapsules develops very rapidly. Some typical examples of such devices are collected in this paragraph, although we have to leave the majority of investigation in this area outside the limited area of our book.

Pseudoephedrine HCl, a highly water-soluble drug, was entrapped within polymeric microspheres by either an oil-in-water (dispersion or co-solvent method) or a water-in-oil-in-water (multiple emulsion method) emulsion-solvent evaporation method. Acceptable drug loading was achieved with the three methods. The microspheres were characterized by dissolution studies, scanning electron microscopy, and differential scanning calorimetry. The microsphere structure or porosity depended strongly on the polymer selected and, to a lesser extent, on the method of preparation. The drug was partly soluble in cellulosic polymers (ethyl cellulose, cellulose acetate butyrate) but insoluble in poly(methyl methacrylate) and poly(D,L-lactide). The microspheres were formulated into an oral suspension dosage form, using concentrated sucrose or sorbitol solutions, glycerol, propylene glycol, or Neobee M-5 oil as suspending vehicles. The amount of drug leached into the storage vehicle leveled off after 2-3 weeks and did not change significantly during further storage. After 6 months, between 77 and 93% of the original drug loading was still present within the microspheres. The storage of the microspheres in the vehicles had little influence on the drug release profiles [38].

Uptake by gut epithelial tissue of 60 nm polystyrene particles was studied in female Sprague-Dawley rats (180 g, 9 weeks old) after 5 days oral dosing by gavage (14 mg/kg). The gut was divided into lymphoid and non-lymphoid tissue of the small and large intestine, prior to analysis for polystyrene by gel permeation chromatography (GPC). Approximately 10% of the administered dose was recovered from the entire GI tract. The total percentage of the administered dose taken up through lymphoid tissue was statistically much greater than through nonlymphoid tissue. It was estimated that 60% of the uptake in the small intestine occurred through the Peyer's patches, even though the patches comprised a small percentage of the total surface area of the small intestinal tissue. A significant amount of the total uptake was shown to occur in the large intestine, particularly in the lymphoid sections of this tissue. These results were confirmed by fluorescence microscopy [39].

Tamada and Langer [40] reviewed the development of the polyanhydrides as bioerodible polymers for drug delivery applications. The topics included design and synthesis of the polymer, physical properties, techniques to fabricate the polymer into drug delivery devices, evaluation of biocompatibility, and example applications of the polyanhydrides. Discussion of the interrelationship between the physical-chemical properties of the polyanhydrides, fabrication methods, and drug release rates was included therein. One section was devoted to a case study to provide a historical perspective of the development a polyanhydride-based drug delivery treatment from the conception of the idea to the final stages of human clinical trials. This section included an outline of the extensive in vitro and in vivo testing that is necessary for development of a new material for biomedical applications.

A method for preparing polyanhydride granules of an injectable size was developed by Tabata et al. [41]. The resulting granules permitted a nearly constant release of low-molecular-weight, water-soluble drugs without an initial burst. The polyanhydrides used were poly(fatty acid dimer), poly(sebacic acid), and their copolymers. The dyes acid orange 63 and p-nitroaniline were used as model compounds for drugs. Polymer degradation and drug release for disks and variously sized granules of copolymers containing drugs, prepared by a water-in-oil (W/O) emulsion method, were compared with those for devices prepared by the usual compression method. In the W/O emulsion method, a mixture of aqueous drug solution and polymer-chloroform solution was emulsified by probe sonication to prepare a very fine W/O emulsion. The powder obtained by freeze-drying of the W/O emulsion was pressed into circular disks. In the compression method, the drug was mechanically mixed with the polymer, and the mixture was compressed into circular disks. The resulting disks were ground to prepare granules of different sizes. The granules encapsulated more than 95% of the drug, irrespective of the preparation method. Both methods were effective in preparing polymer disks capable of controlled drug release without any initial burst. However, as the granule size decreased to an injectable size (diameter, <150 μ), a large difference in the drug-release profile was observed between the two preparation methods. The injectable granules obtained by the W/O emulsion method showed nearly constant drug release without any large initial burst, in contrast to those prepared by the compression method, irrespective of the drug type. Degradation studies of the granules demonstrated no difference in the degradation profile of the granule matrix itself between the two methods. Light microscopic observations of polymer disk prepared by the compression method indicated a nonuniform distribution of dye islands throughout the matrix. In contrast, a highly homogeneous mixing of dye and polymer was achieved for devices prepared by the W/O emulsion method. It is therefore possible that this highly uniform distribution of drug throughout the polymer matrix leads to a reduced initial burst in drug release from the injectable granules obtained by the W/O emulsion method [41].

Mucosal membranes are the most frequent portals of entry of almost all infectious agents. The most important protective humoral factors on mucosal surfaces are locally produced antibodies predominantly of the IgA isotype. Therefore, the induction of specific protective immune responses at mucosal surfaces is a highly desirable goal in the prevention of many infectious diseases. This can be achieved by novel vaccination strategies using effective antigen delivery systems. The ingestion or inhalation of antigens results in the induction of immune responses on mucosal surfaces due to the dissemination of antigen-specific and IgA-committed precursors of plasma cells from intestinal and respiratory lymphoid tissues to other mucosal surfaces. However, only minor quantities of antigens are absorbed from mucosal surfaces due to their degradation by enzymes and hydrochloric acid and inefficient uptake. The protection and enhanced uptake of antigens provided by particles with slow biodegradation could facilitate oral immunization. This possibility was explored in recent studies by Mestecky and colleagues [42], which indicate that simple proteins or complex antigens of viral and bacterial origin incorporated into biodegradable microspheres induce, after ingestion, both mucosal and systemic immune responses. Experimental animals orally immunized with an influenza virus vaccine in biodegradable microspheres displayed virus-specific antibodies in saliva and in serum, and were protected against challenge with the live viruses. Many theoretical and practical advantages of oral over systemic immunization should stimulate further studies of applications of oral vaccines [42].

The transient dynamic swelling and dissolution behavior during the release of a growth hormone releasing peptide, [D-Trp2-D-Phe5]GHRP, from erodible, non-cross-linked poly(methyl methacrylate-co-methacrylic acid) (PMMA/MAA) beads has been investigated at pH = 7.4 as a function of buffer concentration [43]. Although the swelling front penetration showed a ionization-limited behavior similar to that of nonerodible cross-linked PMMA/MAA beads, the normalized diameter of the polymer beads exhibited a brief initial rise followed by an extended linear decline due to establishment of the polymer dissolution process. This was consistent with the general kinetic scheme of dissolution of glassy polymers originally predicted for the slab geometry. In all cases, the initial gel thickness increased as a result of the ionization and swelling of the glassy PMMA/MAA beads. This was followed by an extended period of constant gel thickness due to the onset of polymer dissolution and the synchronization of movement of the swelling and dissolution fronts. The resulting constant gel layer thickness as well as the onset and duration of front synchronization showed an increasing trend with decreasing buffer concentrations. As a result, the corresponding peptide release was slower and the release duration longer at lower buffer concentrations. This was believed to be the first time that a synchronization of swelling and dissolution fronts was documented for a spherical erodible sample. Although such synchronization of fronts did not result in a constant rate of peptide release due to the spherical geometry, some non-Fickian release characteristics were observed [43].

A drug delivery system for biologically active agents targeted to specific cells could be used to improve tissue repair in orthopedics. Such system must be controllable and capable of drug release over an extended period of time [44]. Biodegradable, membrane-moderated, monolithic microspheres for the controlled release of growth hormone (GH) were developed, and the release of GH was monitored in vitro. Cross-linked gelatin microspheres were used as the vehicle, with the drug dispersed within the gelatin. The amount of GH released from the microspheres was increased following ultrasonication. The release of GH was monitored in phosphate buffered saline and horse serum. Interestingly, a higher level of GH was detected in the phosphate buffered saline than in serum. In addition, both pH and enzyme-induced degradation had an effect on the swelling kinetics of the gelatin microspheres. The release of GH from the microspheres was diffusion controlled, during the time period studied [44].

The formulation of a novel controlled release parenteral dosage form was the goal of the investigation of Bayomi and Elsayed [45]. Microbeads of cross-linked casein containing diethylstilboestrol (DEST), as a model drug, were prepared using emulsion polymerization technique. The effects of different concentrations of glutaraldehyde, as a cross-linking agent, on particle size, particle size distribution, shape of microbeads, drug content, as well as the rate of drug release from the microbeads, were then studied. Spherical microbeads with low particle size distribution and high drug load were obtained as glutaraldehyde concentration increased. Furthermore, the release rate from the microbeads was decreasing with the increase of the concentration of the cross-linking agent. The effect of pH of casein solution as well as the amount of added drug on the characteristics of the microbeads were also investigated. The biodegradability of casein microbeads prepared at glutaraldehyde concentration 2.37% was tested in alpha-chymotrypsin solution. The microbeads showed distinct signs of biodegradation within a few days of incubation. No signs of adverse effect were noticed when drug free casein microbeads were injected intraprotenially in mice. It was concluded that casein microbeads could be considered as a good candidate for the preparation of a dependable parenteral time release dosage form [45].

Three different poly[(amino acid ester)phosphazenes] were examined by Allcock et al. [46] to investigate their possible use as drug delivery vehicles. These three polymers were poly[di(ethyl glycinato)phosphazene], poly[di(ethyl alanato)phosphazene] and poly[di(benzyl alanato) phosphazene]. These macromolecules either share the same amino acid residue or the same ester group, and this facilitated comparisons of the hydrolytic decomposition and the small molecule release profiles of the polymers. The polymers were synthesized by treatment of poly(dichlorophosphazene) with an excess of the appropriate amino acid ester. Tetrahydrofuran solutions of each polymer were then thoroughly mixed with ethacrynic acid, a diuretic, or Biebrich Scarlet, an azo dye. Films cast from these solutions were immersed in aqueous media (pH = 7) at 25°C and 37°C for approximately 1,400 hours. During these experiments, the release of the small molecules was monitored by UV/visible spectroscopy [46].

Rational delivery systems for leuprorelin acetate, a potent LHRH agonist, had been achieved by developing a microsphere system using biodegradable polymers, poly(lactic/glycolic acid) (PLGA) and polylactic acid, which sustainedly released the drug depending on the biodegradation of polymer used and persistently suppressed steroidogenesis for over 1 and 3 months, respectively, following a single injection. To produce these systems Okada and colleagues established a novel microencapsulation technique, the in-water drying method, and microspheres with a high trap ratio and small initial burst were obtained [47]. A microsphere could also continuously release the drug for 2 or 4 weeks. Chemoembolization using PLGA microspheres containing an angiogenesis inhibitor, TNP-470, resulted in dramatic regression of VX-2 carcinoma in rabbits. The microsphere system using biodegradable polymers is very useful in designing controlled release delivery and targeted delivery to attain potent and rational therapy [47].

The area of self-assembled fine particulate-based composites (nano composites) was a major thrust in advanced material development. Schnur and coauthors reported on the application of biologically derived, self-assembled cylindrical microstructures to form advanced composite materials for controlled release applications [48]. These microstructures have many applications in the material sciences.

Two biodegradable polymer intermediates, triacrylate-4 and tetraacrylate-7, were prepared by polycondensation reaction of ethyl beta-hydroxybutyrate using glycerol and pentaerythritol as initiators and dibutyltin oxide as a catalyst, followed by functionalization of the hydroxyl end groups with acryloyl chloride in the presence of triethyl amine. These polymer intermediates were characterized using H^1-NMR and IR spectroscopic analysis and were employed as cross-linking agents during polymerization of partially neutralized acrylic acid to obtain the corresponding potentially biodegradable polyacrylates [49].

Biodegradable polyester microspheres, containing stable neutron activatable holmium-165 complexed to acetylacetone (AcAc), were prepared by the solvent evaporation technique and later irradiated in a high neutron flux to convert the Ho-165 to Ho-166, a high energy negatron emitter with excellent therapeutic properties (E_{max} = 1.84 MeV; $t_{1/2}$ = 26.9 hours). Potential applications of the radiotherapeutic spheres include the treatment of hepatic metastases (with 10–40 μ particles) and rheumatoid arthritis (3–10 μ particles). In vitro studies in human plasma showed that optimal irradiated spheres retained >99.6% of the initial Ho-166 activity after 404 hours. In vivo stability studies with administration of irradiated spheres into the portal vein of rabbits (n = 13) revealed that 87.9–97.9% of the initial study activity remained in the rabbit liver after 144 hours. The retentive performance of irradiated polyester spheres was found to be strongly influenced by the polymer type, content of the emulsifying agent, solvent removal conditions, Ho-AcAc concentration, and storage and irradiation conditions of the spheres [50].

Chitosan is a polysaccharide, which has structural characteristics similar to glycosaminoglycans. Chandy and Sharma highlighted its application for controlled drug delivery systems. The steroid drugs, namely testosterone, progesterone, and beta-oestradiol were mixed with chitosan and the films were prepared by evaporation technique. The in vitro release profile of these steroids from the film matrix was monitored, as a function of time, in phosphate buffered saline (PBS, pH = 7.4) at 37°C using a UV-spectrophotometer. The degradation of these chitosan and drug loaded chitosan films was also investigated by weight loss and tensile strength studies. The steroid release from chitosan films was compared with the release of these drugs from their microbeads. It appeared that the films and the microbeads stayed intact during the dissolution study of 90 days and this gave a possibility of using these systems in contraceptive applications [51].

Spheres containing chitosan were prepared by Tapia et al. [52] from a wet mass by extrusion and spheronisation. A polysaccharide was included as a solution in the powder mix. Four formulations were considered, containing two different concentrations of chitosan, each prepared at two different sizes of sphere; either approximately 1 or 2 mm. The release of a model drug (diclofenac sodium) from the spheres was found to be considerably slower than formulations without added chitosan (100% in 6 hours, rather than 30 minutes). Thus it was possible to retard drug release from a sphere, without the need for polymer coating, by use of a hydrophilic gel. The drug-release profiles followed first-order kinetics (for all four systems) and produced a straight line when plotted as a function of the square root of time. A straight line was also obtained when a double logarithmic plot of release as a function of time was prepared, the gradient of which was in the order of 0.5, which would indicate Fickian diffusion, if from a thin slab, but as the release was from a sphere, the process was described as fitting a non-Fickian diffusion model. Dissolution testing at different stirring speeds did not alter the rate of drug release, demonstrating that the diffusion process was controlled within, rather than in the layer of fluid around, the sphere. Thermodynamic activation parameters were calculated and the four formulations were compared by compensation analysis. It was apparent that there was no common mechanism for drug release, but that the concentration of the chitosan related to the enthalpy change, and the Gibbs free energy change correlated with the dissolution rate [52].

Nastruzzi at al. [53] described how the use of different in vitro experimental systems can influence the determination of the drug-release profile from microparticles and the interpretation of the release mechanism. They employed, as model dosage form, the Parlodel LA(R), a recently marketed microsphere system especially designed for bromocriptine-controlled delivery. The release kinetics of bromocriptine from microspheres were determined by using two different experimental approaches: a dialysis method and a flow-through cell method. From the comparison of the obtained data, it clearly appeared that different in vitro experimental models lead to distinct results in terms of drug availability. On the contrary, both series of data can be convincingly fitted with the same mathematical equation, giving almost identical results in terms of postulated release mechanism. Taken together these results indicated that different experimental approaches should always be employed to determine drug-release kinetics from microparticles in order to obtain more reliable information on the therapeutic dose (bioavailable drug, for in vivo experiments) and on the uniformity of different batches of microspheres.

Pavanetto and colleagues prepared corticosteroid-loaded albumin microspheres designed for intra-articular administration [54]. Dexamethasone was chosen as the model drug and bovine serum albumin was used as the biodegradable, natural polymer. Albumin microspheres were produced by spray drying, a "one-step" technique seldom used in the preparation of microparticulate drug delivery systems with particle sizes <10 μ. The effects of both polymer/drug ratio used in the formulations and the different heat-stabilization conditions were evaluated on morphology, size, solubility characteristics, drug loading, and "in vitro" drug release of the microparticles.

Nanospheres with D,L-lactide/glycolide copolymer (PLGA) were prepared by Niwa and colleagues [55] as a biodegradable polymeric carrier for both water-soluble and insoluble drugs by a novel spontaneous emulsification solvent diffusion method. Indomethacin and 5-fluorouracil (5-FU) were employed as poorly water-soluble and water-soluble model drugs, respectively, to investigate the encapsulation efficiency. The drug and PLGA, dissolved in an acetone-dichloromethane (or acetone-chloroform) mixture, were poured into an aqueous solution of polyvinyl alcohol and stirred using a high-speed homogenizer when necessary. The dispersed droplets were finely emulsified into nanometer-sized spheres. The marked decrease of the interfacial tension between organic and aqueous phases and the spontaneous mixing caused by a rapid diffusion of acetone from the organic to aqueous phase resulted in the formation of submicron-sized

PLGA spheres. The recovery of indomethacin entrapped in the nanospheres of 400–600 nm in diameter increased to 75% at maximum. The rapid deposition of polymeric film on the droplet was required for improving the encapsulation of 5-FU to prevent leakage from the droplet. The mean diameter of nanospheres formulated with 5-FU were successfully decreased to 200–300 nm even without high-speed homogenizing. The drug-release behavior from nanospheres suspended in buffered solution exhibited a biphasic pattern. The initial burst of release might be due to the rapid release of drugs deposited on the surface and in the water channels of nanospheres. At a later stage, the drug-release rate was reduced. During the release test, PLGA was not degradated for 100 hours irrespective of the molecular weight. The molecular weight of polymer was a main factor in controlling the drug-release rate from the nanospheres [55].

A comprehesive analysis of poly(D,L-lactide-co-glycolide, 50:50) samples of similar molecular weight was made by Schmitt et al. [56] who obtained this polymer from three commercial sources and characterized it by gel permeation chromatography, differential scanning calorimetry, X-ray powder diffraction, viscometry, and proton nuclear magnetic resonance spectroscopy. Pellets were prepared by melt-pressing spray-dried polymer with a 4-mm standard concave punch and die set and a thermostated holder of original design. Amaranth (5% w/w) was incorporated in pellets used for release studies. Degradation and release studies were conducted at 37°C in pH = 7.2 phosphate buffered saline. The molecular weights of all polymers were found to decrease continuously after exposure to phosphate buffered saline. All polymers showed two distinct regions of molecular weight decrease. Mass loss experiments for all polymers resulted in sigmoidal curves typical of polymers undergoing bulk hydrolysis. The onset of mass loss (defined as 10% mass loss) was found to differ by as much as 6 days among the three polymers studied. The release studies showed an initial burst of release followed by a period of 15–25 days during which little or no dye was released. A second phase of release followed, lasting approximately 10 days, until all dye was released. The time at which release began slightly preceded the onset of mass loss [56].

Poly(DL-lactide/glycolide, 50:50) microspheres containing bovine serum albumin (BSA) were prepared with and without Carbopol-R 951 (a potential adjuvant agent) by oil/oil, oil/water, and (water/oil)/water emulsion methods. The protein loading of the microspheres reached 50–70% of the theoretical amount of protein put into the formulation medium. The microsphere particle size was approximately 500 μ, 25–100 μ, 10–20 μ using oil/oil, oil/water, or (water/oil)/water emulsion techniques, respectively. The release of BSA was dependent on the preparation method. The greatest burst of release was found for vacuum-dried microspheres formulated using the (water/oil)/water method. This burst effect could be eliminated by lyophilizing the microspheres following their preparation. BSA was released at a higher initial rate from microspheres prepared by the oil/water emulsion method that contained Carbopol-R 951 than from microspheres not containing Carbopol-R 951. Release studies also suggested that the release of BSA could be sustained for 54, 36, or 34 days for microspheres prepared by oil/oil, oil/water, or (water/oil)/water methods, respectively [57].

Another application poly(D,L-lactide/glycolide) found was for a fabrication of progesterone-loaded microspheres [58]. Such particles were obtained by a solvent evaporation process from a poly(D,L-lactide-co-glycolide) (85/15 PLG) and from α-progesterone. Methylene chloride was used as solvent, and polyvinyl alcohol and methylcellulose were used as surfactants. The microspheres were characterized by scanning electron microscopy, differential scanning calorimetry, and X-ray powder diagrams. Rosilio et al. showed that the morphology and the thermal behavior of PLG microspheres can vary significantly with progesterone loading and sample thermal history. Below and at 16.5% loading, the microspheres exhibited a smooth outer surface. Above 23% loading, the surface became rough, embedded by copolymer particles or well-defined crystals. Pores and cracks also were observed. Below 35% the progesterone was molecularly dispersed. At 35% and above, crystal domains of the steroid appeared and two crystalline forms were found: α- and β-progesterone. The physical state of progesterone and the nature of its crystal domains dispersed in the PLG matrix can be changed during storage. Also a progressive development of an endothermic peak at the T_g event of the copolymer was observed during storage. However, no well-defined relationship of peak size to progesterone loading was shown [58].

A new method to prepare polyanhydride microspheres capable of near-constant sustained release of low molecular weight, water-soluble molecules was developed by Tabata and Langer [59]. The polyanhydrides used were poly(fatty acid dimer) (PFAD), poly(sebacic acid) (PSA), and their copolymers [P(FAD-SA)]. Acid orange 63 (AO), acid red 8 (AR), and p-nitroaniline, were used as model release molecules. P(FAD-SA) microspheres containing the molecules with or without gelatin were prepared by a modified solvent evaporation method using a double emulsion. The microspheres were spherical with diameters of 50-125 μ and encapsulated more than 85% of the molecule, irrespective of the compound used. Near-zero-order degradation kinetics were observed for 5 days as judged by sebacic acid (SA) release. Microsphere degradation was pH sensitive, being enhanced at high pH, and became more stable in acidic conditions, irrespective of the incorporation of gelatin in the matrix. For the gelatin-free microspheres, a close correlation of SA release and AO release was observed at 2% loading, suggesting a release mechanism that was controlled dominantly by degradation. However, the incorporation of gelatin into the microsphere significantly extended the periods of molecule release from P(FAD-SA) microspheres, although the degradation profile of the microspheres themselves was quite similar to that of gelatin-free microspheres. It is possible that an interaction between FAD monomers and gelatin molecules causes continued release, even after the polymer matrix completely degrades (even after complete degradation, FAD monomers remain because of their poor water solubility). Thermal analysis of polyanhydride microspheres at different degradation stages demonstrated that a crystalline structure was formed between gelatin and the FAD monomers produced with microsphere degradation. This gelatin effect on the extended period of drug release was not observed for microspheres prepared from other polyanhydrides: PSA and its copolymer of bis(p-carboxyphenoxy) propane and SA. It is therefore likely that the crystalline structure formed between gelatin and FAD monomers may function as a reservoir for water-soluble drugs, leading to an extended period of molecule release from the gelatin-loaded P(FAD-SA) microspheres [59].

To develop thermoresponsive microspheres containing antitumor agents, which are expected to embolize the artery or arteriole feeding the tumor region and to release the antitumor drugs pulsatively by the heat in the hyperthermia treatment, biodegradable microspheres consisting of a liquid crystalline and poly(D,L-lactic acid) were prepared by a solvent evaporation method. In vitro studies of the release of model drugs, indomethacin (IND) or aclarubicin (ACR), from the microspheres were performed by changing the temperature between 37 and 43°C (the latter is usually applied in the hyperthermia therapy). Although the differences in ACR release between 37 and 43°C from the microspheres consisting of high-molecular-weight ($M_w = 64,000$) poly(D,L-lactic acid) (HPLA) as a polymer component were observed, the pulsatile release of ACR from the microspheres was not achieved. The effect of an addition of low-molecular-weight ($M_n = 2600$) poly(D,L-lactic acid) (LPLA) as a second polymer component to the microspheres on the pulsatile drug release was evaluated using several microspheres containing IND. It was clarified that the composition of HPLA, LPLA, and the liquid crystalline had an important role in the pulsatile release of IND. The pulsatile release of ACR could be achieved with the microspheres which consisted of HPLA, LPLA, and polyoxyethylene glyceryl tristearate (PGTS) (5/5/2) [60].

Bodmeier et al. [61] formed nanoparticles containing ibuprofen, IND or propranolol after the addition of solutions of the drugs and acrylic polymers (Eudragit RS or RL 100) in the water-miscible solvents, acetone or ethanol, to water without sonication or microfluidization. The colloidal dispersions were stabilized by quaternary ammonium groups and did not require the addition of surfactants or polymeric stabilizers. The nanoparticles were compared to nanoparticles prepared either by a microfluidization-solvent evaporation method with a water-immiscible organic solvent, methylene chloride, or by a melt method with respect to particle size and redispersibility of freeze- or spray-dried samples. Nanoparticles prepared by microfluidization or the melt method were easily redispersed, while Eudragit RS nanoparticles prepared by spontaneous emulsification were not redispersible. Flexible films were formed from the nanosuspensions after the addition of 15% triethyl citrate, a water-soluble plasticizer. The release of propranolol from the films increased with increasing proportion of RL, but was independent of the order of mixing of the two polymers or nanosuspensions during film preparation. The drug release from IND films was increased by adding water-soluble polymers to the nanosuspension [61].

Akbuga [62] varied ratio between Eudragit L:Eudragit RS and Eudragit S:Eudragit RS to alter a controlled release of furosemide from microspheres containing these polymers. A wide range of release rates of drug can be obtained by a simple change in the ratio of polymers. An increase in Eudragit RS content of polymer microsphere matrix brought about a decrease in the release rate. On the other hand, the effect of particle size on the drug release rate from furosemide microspheres was also investigated. The effect of microsphere sizes on release rate depends on the type of Eudragit. The decrease in release rates of small microspheres may be due to agglomerate formation. Dissolution data indicated that the release followed Higuchi's matrix model kinetics.

Supercritical fluids were used by Tom with colleagues [63] to form two different types of microparticles intended for controlled drug-release applications: drug-loaded polymer microspheres and small protein particles. A poly(hydroxyacid), poly(D,L-lactic acid) (DL-PLA) and a pharmaceutical (lovastatin) were dissolved in supercritical CO_2 and coprecipitated by rapid expansion of the resulting supercritical solution (RESS) to form polymer-drug microspheres and microparticles ranging in size from 10 to 100 μ. Variations in the concentration of lovastatin in the precipitate correlated with changes in the precipitate's morphology, ranging from continuous drug-polymer networks to microparticles and/or to microspheres. The formation of polymer-drug microparticles by RESS was assumed to be the first step towards a feasible single-step, low-temperature process that yields solvent and surfactant-free microparticles suitable for controlled drug release. Two model proteins, catalase and insulin, were dissolved in ethanol/water solution and fed continuously and simultaneously with supercritical CO_2 into a crystallizer to precipitate the proteins. The use of supercritical CO_2 as a gas anti-solvent (GAS) produced catalase and insulin

particles ranging from 1 to 5 μ. Particle morphology ranged from microspheres to rectangular-shaped particles and to needles. Micron-sized protein
particles are needed in several controlled release formulations to accom-
modate the high potency and low dosage of such pharmaceuticals and to
achieve a uniform dispersion of the drug in the injectable polymeric mi-
crospherical carrier. GAS crystallization may be, therefore, a potentially
important process for comminution of proteins since conventional particle
reduction methods (spray-drying, lyophilization, milling, grinding) cannot
produce the micron-sized protein particles needed for controlled release of
highly active enzymes.

Levy and Benita [64] evaluated the in vitro release of diazepam from
a submicron oil/water emulsion using the dialysis bag technique diffusion
and the bulk-equilibrium reverse dialysis bag technique. Irrespective of the
nature of the sink solution used, the release rate of diazepam from different emulsion dosage forms remained slow and incomplete as compared to
a diazepam hydroalcoholic solution using the dialysis bag technique. This
was attributed to a marked decrease in the aqueous drug gradient of drug
available for membrane diffusion in the presence of the oily internal phase,
rendering the permeation through the dialysis membrane the rate-limiting
step in the overall kinetic process. It can definitely be deduced that the dialysis bag technique could not be considered an appropriate method to evaluate the true release mechanism of a drug from a colloidal carrier. An in
vitro kinetic model is therefore proposed where the colloidal drug carrier
suspension was directly placed in the release solution and has the opportunity to release the drug content under maximum dilution and perfect sink
conditions. The drug released sampling was performed through immersed
dialysis bags previously filled and equilibrated with the sink solution in the
receptor compartment. The release profiles of diazepam from the actual
submicron emulsion were similar to that observed from marketed aqueous
and emulsion dosage forms correlating well with pharmacokinetic results
reported in the literature. It was found that the release rate from the oily
nanodroplets was faster than the permeation rate through the dialysis membrane which should be the slowest step governing the overall kinetic process
despite rapid and complete diffusion of dissolved drug within less than 1
hour. In view of the overall results it can be concluded that the release of
diazepam from submicron emulsion was found very rapid under perfect
sink conditions [64].

5.1.4. Polymer implants as drug delivery systems

The application of polymeric implants generally requires the accurate study
of their biocompatibility and/or biodegradation rather than a search of spe-

cial mechanisms of a drug release. These mechanisms remain substantially the same as for other kinds of drug delivery systems.

By providing a long-term and localized source of active drug molecules, controlled release polymer implants may reduce the systemic side-effects and dose-to-dose variability associated with conventional drug administration. Implants may be particularly relevant for delivery of drugs to the brain, where therapy is frequently limited by the blood brain barrier [65]. To aid in the design and application of new delivery systems, Saltzman and Radomsky developed methods for modeling drug transport in tissue in the vicinity of a continuous source. Transport was assumed to occur by diffusion with elimination due to irreversible metabolism, reversible binding to fixed tissue components, or partitioning into capillaries. For polymer implants where diffusion in the polymer determines the rate of drug release and the rate of drug release from the polymer, drug concentration in the tissue and drug penetration depend on rates of elimination and diffusion. Qualitatively similar results were obtained for degradable polymers. Model predictions were also used to interpret data on the delivery of the steroid dexamethasone from an ethylene-vinyl acetate copolymer implant in rat brain. General conclusion was drawn that water-soluble, slowly eliminated, and diffusible molecules are the best candidates for polymeric delivery to brain tissues. In contrast to conventional modes of administration, rapid permeation of active molecules through brain capillaries is the most significant barrier to effective drug distribution in the brain [65].

Reports on the controlled release of drugs, including macromolecular drugs, from silicone elastomers and ethylene-vinyl acetate copolymers, based on the formation of channels and cracks in the polymer, were reviewed by Dicolo. Aqueous interconnected pores were produced by osmotically active additives or by using loads of water-soluble drugs exceeding the percolation threshold. The release was generally proportional to the square root of time. Nevertheless, pseudozero-order release kinetics was obtained by adequately controlling the formulation variables. The analysis of the factors controlling the release pattern and rate was followed by in vivo applications of these types of systems [66].

Horak [67] presented a survey of the use of poly(2-hydroxyethyl-methacrylate) (polyHEMA) in ophthalmology, reconstructive and plastic surgery, surgery of burns, and orthopaedy. Hydrogels from polyHEMA are promising for a controlled release of various drugs and for the coating of hemoperfusion sorbents and hemodialysing membranes. They are used for microencapsulation of cells producing insulin. Emboli from polyHEMA are used for the obturation of blood vessels in a number of medical fields. Microparticle carriers are proposed for diagnostic purposes, or also biosensors containing polyHEMA or its copolymers. Various substitutions are made of polyHEMA-nasal and vocal cord implants, replacements of the bone tissue in stomatology, replacements of tendons, prostheses in the cure of laryngial narrowing, and breast prostheses. The main area of use of poly-HEMA is still the production of contact and intraocular lenses. Other biomedical applications described in [67] have not so far found broad practical application.

Permeability of polymer coatings on urea varies greatly with the type of polymer. A conventional test for measuring the effectiveness of coating involves a 7d static dissolution rate of coated urea into concentrated urea solution, but the results are only qualitative. The approach of Zhang et al. [68] was, instead, to make quantitative measurements of permeability, and so make more accurate predictions of release rate of urea across a membrane. A simple device, consisting of a container attached to vertical pipes at the bottom, was constructed to determine permeability of coats on urea granules. A turbulent flow of water ran over a 2 cm pack of coated-urea granules so urea did not accumulate at the outer surface of the coated granules. Separate determinations with two thicknesses of coats (8.8 and 14.7 μ) were conducted with water at 12 or 31°C. Then permeability and activation energy of permeability were calculated. A comparison was also made between release rate of urea calculated from permeability and that determined by 7d dissolution rate method at 23°C. Nearly 100 hours were required for 100% release with the thick coating, but only 20% urea was released after 168 hours with the 7d dissolution rate method [68].

The influence on blood of polyurethaneurea hydrogels in vitro was investigated by Yu et al. based on poly(ethylene oxide). A hydrogel was compared with the regenerated cellulose membrane Cuprophan in terms of complement activation. The hydrogel induced less complement activation and the presence of poly(ethylene oxide) is likely to be beneficial to platelet reactivity. The ability to vary the polymer composition and the solubility of the polymers in organic solvents makes the polyurethaneurea hydrogels strong candidates for composite biomaterials [69].

Hashizoe and colleagues designed a new device, a scleral plug, that releases drugs into the vitreous after being implanted and fixed at the pars plana. Use of the plug for provision of doxorubicin hydrochloride was evaluated in rabbits. The scleral plug (8.5 mg) was made of poly(lactic-glycolic acid) (molecular weight was 40,000 daltons) containing 1% doxorubicin. Vitreous concentrations of doxorubicin were measured after the implantation. In vitro studies showed that the plug released 26% of the drug during 4 weeks. In vivo studies demonstrated that the concentration in the vitreous humor was maintained at a therapeutic range for longer than 4 weeks. No substantial toxic reactions were observed by electroretinographic and histopathologic evaluations. The findings suggested that a scleral plug made of biodegradable polymers is a promising device for a controlled drug-release system in the vitreous [70].

Prolonged reversible nerve blockade has broad applications in a number of clinical areas involving acute or chronic pain. The desired periods of reversible nerve blockade could vary from as little as one day to as long as one week. Implantation of a biodegradable controlled-release local anesthetic device adjacent to nerves could potentially be a valuable dosage form. Controlled release devices based on polyanhydride polymers were prepared by Maniar et al. [71] who studied a release of the local anesthetic bupivacaine HCl with varying rates. The parameters affecting the release of drug were studied to optimize the formulation. The studies were conducted with rectangular devices consisting of bupivacaine HCl dispersed homogeneously in the polymer at a loading of 10% w/w. Devices fabricated from three different copolymers, synthesized from fatty acid dimer and sebacic acid, were studied to determine the effect of comonomer on the release kinetics of the drug. Release studies were conducted at pH 7.4, 37ºC, and the release profiles were analyzed to determine the mechanism of release. The release of bupivacaine HCl could be best described by first-order release kinetics from all three copolymers and the release rate constant, k_r, was directly proportional to the hydrophilicity of the polymer. The first-order release rate constants were linearly proportional to both the erosion rate and drug release rate ($r_2 = 0.999$). Release profiles from all three copolymers could also be described by an equation derived for a surface eroding cylindrical device. The erosion rate was obtained by fitting the release profile to the equation using a nonlinear regression method. The results showed that the drug release was controlled by erosion for the three copolymers, P(FAD-SA) 30:70, 20:80, and 10:90, and the release rates were 0.0004, 0.00066, and 0.0012 g/(cm^2*day), respectively. In addition, release profiles expressed as M_t/M_{oo} (fractional agent release profile) against t/t_{oo} fit the theoretical equation for all three copolymers. These results suggest that polyanhydrides undergo pure surface erosion at pH = 7.4 and, therefore, the device geometry and erosion rate determine the release kinetics. Thus, knowing the erosion rate of the fatty acid dimer based polyanhydride, would help in achieving the appropriate drug-release kinetics by manipulating the geometry of the device [71].

Following the identification of antibodies (Abs) as agents of immunity, it was hypothesized that individuals could be both (1) protected against disease by the transfer of unmodified Abs (passive immunization), and (2) cured of established disease by Abs armed with cytotoxic agents (immunotherapy). The development of monoclonal antibody (mAb) technology in 1975 reinvigorated these ideas. Although passive immunization has

been practiced with great success for many years, successful tissue targeting by systematically delivered immunotoxins in humans has been documented in only a few cases. New modes of drug delivery, engineered for
mAb-based products, may enable new applications of passive immunization and may provide improved tissue targeting for immunotherapy. By allowing sustained and tissue-localized delivery of mAb-conjugates, controlled-release polymers may play an important role in this effort. Saltzman
[72] reviewed the use of mAb in treating and preventing human disease,
as well as the pharmacokinetics of Ab delivery. Two areas where controlled
Ab release may yield new therapies were highlighted therein to be sustained passive immunization of the mucus secretions and immunotherapy
of brain tumors.

The synthesis, characterization, and reactivity of alternating copolymers of maleic anhydride (MA) with new vinyl ethers of mono-O-methyl
oligoethylene glycol (MePegnVE, degree of oligomerization n = 1–4) were
described in [73]. The linear MA/PegnVE copolymers were partially es-
terified with alcohols such as methanol, ethanol, n-butanol, methoxy-
ethanol, and n-dodecanol. Depending upon the nature of the two reactive
components, the degree of esterification was between 10 and 70%. The potential of this new class of functional polymers is to provide amphiphilic
polymeric ionomers as candidates for drug controlled release, based on
their biocompatibility [73].

Fibrous materials always play an important role in the advancement
of medicine and surgery. Their uses range from nonimplantables to extracorporeal devices and implants. Some examples of implants and the extracorporeal devices are discussed in [74, 75] to show how the past is guid-
ing us to the future.

5.2. SYSTEMS OF CONTROLLED RELEASE IN
AGRICULTURE AND INDUSTRY

Systems of controlled release in agriculture and industry work on the basis of the same principles as the systems applied in medicine. They have,
however, some specific features which are mostly defined by the medium
where these systems are used.

Encapsulation techniques are used in the food and cosmetic industries
both to control the delivery of encapsulated agents as well as to protect
those agents from environmental degradation. In foods, the most important
applications of encapsulation include: (i) encapsulation of flavors, (ii)
shielding of oxygen- or water-sensitive components such as vitamins, and
(iii) isolation of reactive ingredients until their release is desired. Along
similar lines, the cosmetics industry utilizes encapsulation to: (i) trap

fragrances for controlled or sustained release, (ii) protect particularly volatile components, and (iii) provide release at a delayed time. The preparation methods for these microcapsules may be similar, whether they contain flavors, fragrances, or drugs. Brannon-Peppas recently presented an overview [76] of the uses of microcapsules in the cosmetics and food industries, with an emphasis on the materials used and the goals and uses of the final products.

Performance characteristics of polyethylene tube dispensers containing a mixture of (E,E)-8,10-dodecadien-1-ol (1), dodecan-1-ol (2), and tetradecan-1-ol (3) were evaluated for suitability as a mating disruptant for codling moth control. The rate of loss of pheromone component from a dispenser at any time was found to be described by the equation: $-dP/dt = (k_1k_2 + k_d)P$, where P is the amount of pheromone component in the dispenser wall; k_1 is the ratio of the amount dissolved in the dispenser wall to the total amount in the dispenser wall; k_2 is the ratio of the evaporation rate to the amount dissolved in the dispenser wall; k_d is the rate constant for chemical decomposition. The evaporation rate, E, of a pheromone component at any time was given by: $E = k_1k_2P$. For all three components during the first three weeks, k_1 decreased from 0.25 to 0.10 and was approximately constant thereafter. The decrease of k_1 with time may have been caused by weather-induced cross-linking of the polyethylene. Over time, k_2 was constant and was 1.27×10^{-3} hour^{-1} for 1, 1.96×10^{-3} hour^{-1} for 2, and 0.31×10^{-3} hour^{-1} for 3. The k_d was zero for 2 and 3 and 6.96×10^{-4} for 1. After 150 days in an orchard in 1991, 95 % of 1 was lost from the dispensers (61 % of the loss was by chemical decomposition and 39% by evaporation). The heat summation units in a Yakima valley orchard during 1991 were 4.7% above the average for the 1980–1991 period, while during 1990 they were the highest for this period (26% above average). After the first three weeks of dispenser aging, the regression line half-lives for 1 for 1990 and 1991 were 31.0 and 35.1 days, respectively. The difference in temperature between 1990 and 1991 did not affect the half-life of 1 very much because so much of the loss was from photochemically induced decomposition. Based on an estimate of the required minimum evaporation rate for mating disruption of 2 mg/(ha∗hour) and a half-life of 35 days for 1, 2,345 dispensers/ha would be required for one application per season; 944 dispensers/ha for two applications per season; and 734 dispensers/ha for three applications per season. If a different emission rate of 1 was required for reliable mating disruption, then the number of dispensers required would be changed proportionately [77].

Controlled-release N fertilizers are commonly used in the production of container-grown ornamental crops, yet the relative effects of various nutrient sources on N leaching are not well known. A 27-week experiment was conducted by Mikkelsen and colleagues [78] to evaluate N leaching loss and plant growth following two applications of six controlled-release N fertilizers and one soluble N fertilizer to container-grown Euonymus patens Rehd. The controlled-release fertilizers evaluated were (noncoated) isobutylidene diurea, oxamide, urea formaldehyde, and (coated) Osmocote, Prokote Plus, and sulfur-coated urea. Of the fertilizers tested, the coated fertilizers generally out-performed the noncoated fertilizers in reducing N leaching losses, stimulating plant growth, and increasing tissue N concentrations. Low N concentrations in the leachate of some treatments indicated efficient nutrient use by the plant. In other treatments, low N concentrations in the leachate merely reflected incomplete N release from the fertilizer. A daily application of NH_4NO_3 resulted in a constant rate of N loss but was not the most effective in promoting growth. Plant growth, tissue N concentrations, and N leaching losses were all increased by doubling the fertilizer application rate from 1 kg N/m^3 to 2 kg N/m^3 [78].

Controlled-release formulations of fenthion (10 and 20%) were prepared with sodium carboxymethylcellulose and an interactive polymer, gelatin below its isoelectric point followed by cross-linking with cupric ion. The formulations were analyzed in the laboratory to monitor the physical integrity of the matrices and release profile of fenthion in water. The matrix stability up to 39 weeks with the cross-linked carboxymethylcellulose matrices was enhanced to 55 weeks when gelatin was incorporated. The average release rate and apparent diffusion coefficient of fenthion from these matrices fluctuated between 5.15 and 11.91 mg/week and 4.00×10^{-9} and 2.01×10^{-3} cm^2/s, respectively. The analysis of the release profiles indicated that the release pattern of fenthion from cross-linked carboxymethylcellulose matrices exhibited a Fickian release pattern, and those that contained gelatin exhibited non-Fickian release characteristics. These biodegradable monolithic formulations are assumed to be applied in mosquito control programs with reduced environmental impact [79].

Controlled-release sources of N and K fertilizers were compared with soluble sources on young "Valencia" orange trees. The effects of these fertilizers on leaf mineral concentration, soil chemical analysis, and tree growth were evaluated by Zekri and Koo for 3 years. Soluble fertilizers were generally more readily available but had shorter residual effects on leaves and soil than controlled-release fertilizers. In the top 30 cm of soil, the plots treated with controlled-release N had 23% more total N than those treated with soluble N sources, while the plots fertilized with controlled-release K contained 56% more extractable K than those that received soluble K. Different effects on leaf and soil N between the two controlled-release N sources, sulfur-coated urea (SCU) and methylene urea (MU), were also found. With the use of controlled-release fertilizers, application frequency was reduced from a total of 15 to 6 applications with no adverse effects on tree growth, leaf mineral composition, or soil fertility during the first 3 years. Combining soluble and controlled-release fertilizers in a plant nutrition program offers an economical and effective strategy for citrus growers [80].

A potential application for using styrene butadiene rubber formulations in the controlled release of nitogen fertilizer in agriculture was applied to tomato plants [81]. The slow release technique of urea and ammonium nitrate proved to be more effective than the conventional method; tomato growth increased vigorously, the hormonal balance equilibrium was enhanced, the yield increased, and fruit quality improved. The application of this technique overcame the drawbacks of using ammonium nitrate observed using the conventional method.

Leaf N and soil nitrate and ammonium levels were monitored in 1986. and 1987 [82] following N fertilization of 8–9 year old highbush blueberries. Urea was applied at 76 kg N/ha in a single application at bud break or in two applications (split) at bud break and petal fall. Controlled-release fertilizers of two different residual effects were applied at 38 kg N/ha or 76 kg N/ha at bud break. Compared to controls, N applications increased soil ammonium and nitrate levels early in the season and leaf N levels throughout the season. Urea provided a greater increase in leaf N and soil ammonium levels than controlled-release fertilizers. Split urea applications increase leaf levels slightly over single urea treatments. Fertilizers increased soil ammonium and nitrate levels below the root zone, indicating that some leaching losses occurred.

The polymeric formulations of plant growth regulators (PGRs) are high molecular weight systems in which the PGR unit is attached to the polymeric chain by a hydrolyzable chemical bond. These polymeric derivatives testers, (ethers, or else) of PGRs are characterized by the ability to release the active compound (PGR) from their solutions (mainly aqueous) in certain conditions. The release of the PGR can be controlled by external factors (pH, temperature, enzymes, solution concentration) and inherent properties of the whole macrosystem chemical structure, such as the type of the hydrolyzable bond between PGR unit and the main polymeric chain, the structure of the polymer chain (e.g., molecular weight, level of hydrophilicity, and the content of hydrophobic groups: macromolecular conformation in solution, etc.). These controlled- (slow) release PGRs display certain advantages over conventional PGR formulations due to their prolonged action, improved efficiency (e.g., wide range of effective concentrations), and greater safety to nontarget organisms and the applicators. In addition, the ability to alter the solubility level and modify the application form is of considerable interest. The biological activity efficiency of polymeric PGRs was documented and the relation of this efficiency to the PGR unit hydrolytic release ability was mentioned by Tsatsakis and Shtilman [83]. Slow-release polymeric PGRs were considered therein to solve certain problems in agriculture.

Trials for the control of soil pests, particularly of termites (Isoptera-Termitidae), in groundnuts (Arachis hypogaea) in India and Sudan used chlorpyrifos and isofenphos granules, chlorpyrifos, phorate, carbosulfan, and carbofuran in controlled-release formulations, and chlorpyrifos seed dressing [84]. Their effects on foliar pests were also noted in that paper. Chlorpyrifos controlled release pellets were as effective as aldrin, used as a standard, in reducing root and pod attack and, like aldrin, doubled yields. Isofenphos and chlorpyrifos granules increased yields and reduced pod damage, but to a lesser extent. Other treatments were less effective. Carbosulfan and phorate controlled release and isofenphos granules reduced leaf miner attack. These trials establish the efficacy of controlling termites and other soil pests with controlled-release formulations of otherwise non-persistent insecticides. However, the expensive formulation is unlikely to be cost-effective for rural farmers in developing countries and, in the case of chlorpyrifos, residue levels in kernels may be unacceptable. The authors plan to extend their work by investigation of other insecticides in the formulation and development of cheaper controlled release matrices [84].

St-UF, starch cross-linked by urea-formaldehyde prepolymer, was established [85] as an effective matrix for controlled release of the systemic pesticide carbofuran for broadcast application to soil against various pests in agriculture. The water uptake and carbofuran release of this system fits in with the generalized equation $M_t/M_{oo} = kt^n$ applicable to controlled release systems involving swelling. The St-UF matrix showed an inverse relationship between rate of release and extent of cross-linking. The release rate at a particular cross-linking value was not affected by the extent of loading at low levels (3–10%), but increased at higher loading levels (20%) due to increase in porosity. The release rates were studied in mixtures of water and methanol at 0, 50, 70, and 100% methanol levels, thereby varying the solubility of carbofuran in the medium from 1.2×10^3 to 100×10^3 ppm. The release rates increased with solubility in a linear manner, indicating the porous nature of the matrix. Single particle release studies were conducted by a novel procedure for the first time for a matrix dispersed system of irregular shape. The results established that the mechanism of release was non-Fickian.

Tuber dormancy can be released immediately in many commercially important potato cultivars by brief treatment (1–2 days) with bromoethane (BE) vapor at room temperature [86]. The development of a large scale technology for BE application and safe removal through a capturing technique is necessary for successful application of this dormancy release method. Ideally, BE treatment of seed tubers would occur in a closed environment that would capture BE vapor in an unaltered form and allow controlled release for treatment of subsequent tuber lots. Results of screening studies for adsorbents indicated that the medium capacity activated carbon adsorbent has (i) a high capacity for BE; (ii) a low capacity for water; and, (iii) adsorbs and deadsorbs BE quickly and easily. A plausible design of a large scale, dormancy release facility was presented on the basis of the aforesaid results. The proposed facility should meet present goals of the seed potato industry in an environmentally responsible manner [86].

A significant amount of residual monomer or short chain polymers remain unbound in set composite material. Due to its potential impact on both the biocompatibility and the structural stability of the restoration, many investigators have studied the elution of these unbound molecules into aqueous media. The results of these studies suggest that elution of leachable components from composites is rapid, with the majority being released within a matter of hours. Weight losses of up to 2% of the mass of the composite were reported in [87] under certain conditions. The studies showed that the extent and rate of elution of components from composites is dependent upon several factors. The quantity of leachables was correlated to the degree of cure of the polymer network. The composition and solubility characteristics of the extraction solvent influence the kinetics and mechanism of the elution process. Elution was generally thought to occur via diffusion of molecules through the resin matrix and was therefore dependent upon the size and chemical characteristics of the leachable species [87].

The effects of incorporating a hydrogel polymer into sand on the development of selected horticultural plants grown under saline conditions was demonstrated by Elsayed, Kirkwood, and Graham [88]. In separate experiments, the seeds of tomato (Lycopersicon esculentum Mill.), lettuce (Lactuca sativa L.), and cucumber (Cucumis sativus L.) were germinated in sand-swollen hydrogel polymer mixture (25:75, by volume) with added Hoagland nutrient solution. At cotyledon + first true leaf stage, the plantlets were transplanted into polythene growbags containing a range of sand-swollen hydrogel polymer combinations (0:100, 25:75, 50:50, 75:25, and 100:0, by volume). Saline solutions containing NaCl, $CaCl_2$, and $MgCl_2$ were prepared as molar solutions and applied at combined concentrations as follows; Control (Hoagland), 2,000, 4,000, 8,000, and 32,000 ppm. Application of the appropriate solution to the growbags was made twice per week, alternating with a comparable watering regime. Harvesting was carried out after 14 and 28 days. Polymer incorporation encouraged growth of all species under all saline conditions, the order of effectiveness of the polymer contents being as follows; 75% > 50% > 25% > 10% > 0%. At high salinity (32,000 ppm) plants of the test species were reduced in growth but appeared to be tolerant at all levels of polymer incorporation; in pure sand the level of tolerance in tomato and cucumber was < 8,000 ppm and in lettuce $< 4{,}000$ ppm. Generally, dry weight, leaf area, succulence, chloroplast pigments (chlorophyll a, chlorophyll b, and carotenoids), photosynthetic activity, total amino acids, proline, and protein contents were increased with polymer incorporation compared with pure sand. This hydrogel polymer appears to be highly effective for use as a soil conditioner in horticulture, to improve crop tolerance and growth in a sand or light gravel substrate under saline conditions [88].

A system of partial differential equations was developed by Smith and Peppas [89] to describe the transient, three-step transport of moisture through packaged food products. The steps included (1) Fickian diffusion through the polymer package film; (2) Langmuirian adsorption upon the food surface; and (3) Fickian diffusion through the food material. A set of finite difference equations was derived to approximate the continuous model. These equations were solved for standard boundary conditions in each section of the food packages. The results can be used for the determination of the food package shelf-life [89].

A novel controlled-release technology based on side-chain crystallizable polymers was developed by Greene et al. for further application of these polymers in agriculture and medicine. Laboratory and field data which are presented in [90] confirmed that the release of active ingredients from microcapsule formulations of pesticides can be triggered by increasing soil temperature. This can result in a reduction in pesticide application rates.

Controlled release of iodine in water is of great interest in some developing countries for people's health. A new way was explored by using a natural and common material such as wood [91]. The kinetics of absorption by wood of iodine dissolved in methanol was first studied, as well as the following stage of evaporation of methanol. The wood containing iodine was thus immersed in water, and the kinetics of release of iodine was followed. The process of absorption and desorption was controlled by diffusion with a coefficient of matter transfer on the surface. A numerical model based on finite differences was successfully tested for describing the process [91].

The widescale use of bioactive agents has brought many advantages, particularly to the agricultural industry. However, there are increasing worries about further escalation of the use of bioactive species, in view of the associated problems, e.g., environmental pollution, accompanying their use. The development and application of controlled-release technology has held out the promise of solving many of these problems. The technologies which involve delivery systems in which the agrochemical is chemically bound to a polymer have been reviewed by Kenawy and coauthors in [92]. Since the strategy and chemistry involved in polymeric anti-fouling paints and in polymeric wood preservatives are very similar to those adopted with agrochemicals, short sections on these topics were also included therein [92].

REFERENCES

[1] K. Petrak. *British Polym. J.,* **22**, 213 (1990).

[2] J. Kost. *Polymer News,* **20**, 73 (1995).

[3] R.L. Dunn. *ACS Symp. Ser.,* **469**, 11 (1991).

[4] I.C. Jacobs, N.S. Mason. *ACS Symp. Ser.,* **520**, 1 (1993).

[5] A.L. Iordanskii, T.E. Rudakova, G.E. Zaikov. *Interaction of Polymers with Bioactive and Corrosive Media,* VSP, Utrecht, The Netherlands (1994).

[6] P. Giunchedi, U.Conte, L. Maggi, A. Lamanna. *Intern. J. Pharm.,* **85**, 141 (1992).

[7] R.A. Siegel et al. *J. Controlled Release,* **8**, 179 (1988).

[8] S.B. Mitra. In: *Polymers as Biomaterials,* S.W. Shalaby et al. (Eds.), p. 293, Plenum Press, New York (1984).

[9] V. Vigoreaux, E.S. Ghaly. *Drug Develop. Industr. Pharm.,* **20**, 2519 (1994).

[10] G.M. Zentner, G.A. McCleland, S.C. Sutton. *J. Controlled Release,* **16**, 237 (1991).

[11] G.A. McCleland, S.C. Sutton, K. Engle, G.M. Zentner. *Pharm. Res.,* **8**, 88 (1991).

[12] J.M. Vergnand. *Intern. J. Pharm.,* **90**, 89 (1993).

[13] M. Kelbert, S.R. Bechard. *Drug Develop. Industr. Pharm.,* **18**, 519 (1992).

[14] M.P. Dankwerts. *Intern. J. Pharm.,* **112**, 37 (1994).

[15] F. Pozzi et al. *J. Controlled Release,* **31**, 99 (1994).

[16] K.P Shah, L. Chafetz. *Intern. J. Pharm.,* **109**, 271 (1994).

[17] F.M. Hashem, M.H. Elshaboury, K.E. Gabr. *Drug Develop. Industr. Pharm.,* **20**, 1795 (1994).

[18] M. Otsuka, Y. Matsuda. *Pharm. Res.,* **11**, 351 (1994).

[19] H.R. Bhagat et al. *Drug Develop. Industr. Pharm.,* **20**, 387 (1994).

[20] T. Ostberg, C. Graffner. *Acta Pharmaceutica Nordica,* **4**, 201 (1992).

[21] T.L. Calvert, R.J. Phillips, S.R. Dunkan. *AICHE J.,* **40**, 1449 (1994).

[22] K.H. Yuen, A.A. Desmukh, J.M. Newton. *Drug Develop. Industr. Pharm.,* **19**, 855 (1993).

[23] K.H. Yuen, A.A. Desmukh, J.M. Newton. *Pharm. Res.,* **10**, 588 (1993).

[24] R.S.P. Matharu, J.K. Lalla. *Drug Develop. Industr. Pharm.,* **20**, 1225 (1994).

[25] B. Ramaaj, G. Radhakrishnan. *J. Appl. Polym. Sci.,* **51**, 979 (1994).

[26] Y. Kawashima et al. *Pharm. Res.,* **10**, 351 (1993).

[27] Y. Kawashima et al. *Chem. Pharm. Bull.,* **41**, 2156 (1993).

[28] P.I. Lee *J. Pharm. Sci.,* **82**, 964 (1993).

[29] E.J. Vanhoogdalem, A.G. Deboer, D.D. Breimer. *Clinical Pharmacokinetics,* **21**, 11 (1991).

[30] Y.W. Chien. *Drug Develop. Industr. Pharm.,* **20**, 417 (1994).
[31] H.Ho, Y.W. Chien. *Drug Develop. Industr. Pharm.,* **19**, 295 (1993).
[32] B. Berner, V.A. John. *Clinical Pharmokinetics,* **26**, 121 (1994).
[33] S.K. Jain, S.P. Vyas, V.K. Dixit. *Drug Develop. Industr. Pharm.,* **20**, 1991 (1994).
[34] M.J. Pikal. *Adv. Drug Deliv. Rev.,* **9**, 201 (1992).
[35] A. Gopferich, G. Lee. *Intern. J. Pharm.,* **71**, 237 (1991).
[36] R. Toddywala, Y.W. Chien. *Drug Develop. Industr. Pharm.,* **17**, 245 (1991).
[37] R. Burrows, J.H. Collett, D. Attwood. *Intern. J. Pharm.,* **111**, 283 (1994).
[38] R. Bodmeier, H. Chen, P. Tyle, P. Jarosz. *J. Controlled Release,* **15**, 65 (1991).
[39] A.M. Hillery, P.U. Jani, A.T. Florence. *J. Drug Targeting,* **2**, 151 (1994).
[40] J. Tamada, R. Langer. *J. Biomater. Sci. Polym. Ed.,* **3**, 315 (1992).
[41] Y. Tabata, A. Domb, R. Langer. *J. Pharm. Sci.,* **83**, 5 (1994).
[42] J. Mestecky et al. *J. Controlled Release,* **28**, 131 (1994).
[43] P.I. Lee. *Pharm. Res.,* **10**, 980 (1993).
[44] L. Disilvio et al. *Biomaterials,* **15**, 931 (1994).
[45] M.A. Bayomi, Y.M. Elsayed. *Drug Develop. Industr. Pharm.,* **20**, 2607 (1994).
[46] H.R. Allcock, S.R. Pucher, A.G. Scopelianos. *Biomaterials,* **15**, 563 (1994).
[47] H. Okada et al. *J. Controlled Release,* **28**, 121 (1994).
[48] J.M. Schnur, R. Price, A.S. Rudolph. *J. Controlled Release,* **28**, 3 (1994).
[49] A.B. Argade, N.A. Peppas. *Polymer Bull.,* **31**, 401 (1993).
[50] R.J. Mumper, M. Jay. *J. Controlled Release,* **18**, 193 (1992).
[51] T. Chandy, C.P. Sharma. *Biomater. Artificial cells and Immobilization Biotech.,* **19**, 745 (1991).
[52] C. Tapia, G. Buckton, J.M. Newton. *Intern. J. Pharm.,* **92**, 211 (1993).
[53] C. Nastruzzi et al. *J. Microencapsulation,* **11**, 565 (1994).
[54] F. Pavanetto et al. *J. Microencapsulation,* **11**, 445 (1994).
[55] T. Niwa et al. *J. Controlled Release,* **25**, 89 (1993).
[56] E.A. Schmitt, D.R. Flanagan, R.J. Linhardt. *J. Pharm. Sci.,* **82**, 326 (1993).
[57] H.T. Wang, E.A. Schmitt, D.R. Flanagan, R.J. Linhardt. *J. Controlled Release,* **17**, 23 (1991).
[58] V. Rosilio et al. *J. Biomed. Mater. Res.,* **25**, 667 (1991).
[59] Y. Tabata, R. Langer. *Pharm. Res.,* **10**, 391 (1993).
[60] I. Nozawa et al. *J. Controlled Release,* **17**, 33 (1991).
[61] R. Bodmeier, H. Chen, P. Tyle, P. Jarosz. *J. Microencapsulation,* **8**, 161 (1991).
[62] J. Akbuga. *Drug Develop. Industr. Pharm.,* **17**, 593 (1991).
[63] J.W. Tom et al. *ACS Symp. Ser.,* **514**, 238 (1993).
[64] M.Y. Levy, S. Benita. *Intern. J. Pharm.,* **66**, 29 (1990).
[65] W.M. Saltzman, M.L. Radomsky. *Chem. Eng. Sci.,* **46**, 2429 (1991).
[66] G. Dicolo. *Biomaterials,* **13**, 850 (1992).
[67] D. Horak. *Chemicke Listy,* **86**, 681 (1992).
[68] M.C. Zhang, M. Nyborg, J.T. Ryan. *Fertilizer Res.,* **38**, 47 (1994).
[69] J. Yu et al. *Biomaterials,* **12**, 119 (1991).
[70] M. Hashizoe. *Archives Ophtalmol.,* **112**, 1380 (1994).
[71] M. Maniar, A. Domb, A. Haffer, J. Shah. *J. Controlled Release,* **30**, 233 (1994).
[72] W.M. Saltzman. *Critical Rev. Therap. Drug Carrier Systems,* **10**, 111 (1993).
[73] E. Chiellini et al. *J. Bioactive and Compatible Polymers,* **7**, 161 (1992).
[74] D.J. Lyman. *ACS Symp. Ser.,* **457**, 116 (1991).
[75] D.J. Lyman. *Chemtech,* **23**, 42 (1993).
[76] L. Brannon-Peppas. *ACS Symp. Ser.,* **520**, 42 (1993).
[77] L.M. McDonough, W.C. Aller, A.L. Knight. *J. Chem. Ecology,* **18**, 2177 (1992).
[78] R.L. Mikkelsen, H.M. Williams, A.D. Behel. *Fertilizer Res.* **37**, 43 (1994).
[79] M.P. Prasard, M. Kalyanasundaram. *J. Controlled Release,* **27**, 219 (1993).
[80] M. Zekri, R.C.J. Koo. *J. Amer. Soc. Horticultural Sci.,* **116**, 987 (1991).

[81] F.M. Helaly. *Plastics Rubber and Composites Processing and Applications*, **15**, 125 (1991).
[82] J.B. Retamales, E.J. Hanson. *Communications in Soil Science and Plant Analysis*, **21**, 2067 (1990).
[83] A.M. Tsatsakis, M.I. Shtilman. *Plant Growth Regulation*, **14**, 69 (1994).
[84] J.W.M. Logan et al. *Bull. Entomol. Res.*, **82**, 57 (1992).
[85] P.G. Shukla et al. *J. Controlled Release*, **15**, 153 (1991).
[86] W.K. Coleman et al. *Amer. Potato J.* **69**, 437 (1992).
[87] J.L. Ferracane. *J. Oral Rehabilitation*, **21**, 441 (1994).
[88] H. Elsayed, R.C. Kirkwood, N.B. Graham. *J. Experim. Botany*, **42**, 891 (1991).
[89] J.S. Smith, N.A. Peppas. *J. Appl. Polym. Sci.*, **43**, 1219 (1991).
[90] L. Greene, L.X. Phan, E.E. Schmitt, J.M. Mohr. *ACS Symp. Ser.*, **520**, 244 (1993).
[91] B. Fakhouri, J.M. Vergnaud. *Holzforschung*, **48**, 49 (1994).
[92] E.R. Kenawy, D.C. Sherrington, A. Akelah. *Eur. Polym. J.*, **28**, 841 (1992).

CONCLUDING REMARKS

Since the 1970s polymeric controlled delivery of drugs and other low-molecular weight compounds has become an important area of research, development, and production. In this short time, a number of devices with programmed kinetics and rate of release have progressed from the university or institution laboratory to the hospitals and agricultural fields, and the shops.

While newer and more specific polymeric controlled delivery systems continue to be developed, the understanding of the mechanism of multicomponent transport becomes more and more important for the design of particular devices. The theories discussed in this book represent fundamental approaches which were developed to describe the regularities of multicomponent transport in polymers. Being related to the systems for controlled release, these regularities may be used to design devices more properly corresponding to the practical requirements. The successful application of fundamental research will be a significant challenge which may provide new acceleration of the investigations in this an important and fruitful area of our life.

ACKNOWLEDGMENT

The authors thank their colleagues in the Laboratory of Chemical Stability of Polymers of the Institute of Chemical Physics, Moscow, for their assistance, in particular Dr. N.N. Madyuskin and Ms. L.A. Zimina. We address our special thanks to Dr. Ian C. McNeill, Chemistry Department, the University of Glasgow, Scotland, UK, and Dr. J.H. Petropoulos, "Demokritos," National Research Center for Physical Sciences, Athens, Greece, for the excellent opportunity to work in their laboratories and their very useful comments. We also thank Mrs. A.I. Shalnova, Intergraph, Moscow, for her help in the preparation of this book. And we wish to thank very much our families who do their best to support our work at this uneasy time for Russian science.

SUBJECT INDEX